Lecture Notes in Artificial Intelligence   3601

Edited by J. G. Carbonell and J. Siekmann

Subseries of Lecture Notes in Computer Science

Gianluca Moro   Sonia Bergamaschi
Karl Aberer (Eds.)

Agents
and Peer-to-Peer
Computing

Third International Workshop, AP2PC 2004
New York, NY, USA, July 19, 2004
Revised and Invited Papers

 Springer

Series Editors

Jaime G. Carbonell, Carnegie Mellon University, Pittsburgh, PA, USA
Jörg Siekmann, University of Saarland, Saarbrücken, Germany

Volume Editors

Gianluca Moro
University of Bologna
Department of Electronics, Computer Science and Systems
Via Venezia, 52, 47023 Cesena (FC), Italy
E-mail: gmoro@deis.unibo.it

Sonia Bergamaschi
University of Modena and Reggio Emilia
IEIIT CNR, Dip. di Ingegneria dell'Informazione
via Vignolese, 905, 41100 Modena, Italy
E-mail: bergamaschi.sonia@unimo.it

Karl Aberer
Ecole Polytechnique Fédérale de Lausanne (EPFL)
School of Computer and Communication Sciences
1015 Lausanne, Switzerland
E-mail: karl.aberer@epfl.ch

Library of Congress Control Number: 2005934716

CR Subject Classification (1998): I.2.11, I.2, C.2.4, C.2, H.4, H.3, K.4.4

ISSN   0302-9743
ISBN-10   3-540-29755-3 Springer Berlin Heidelberg New York
ISBN-13   978-3-540-29755-0 Springer Berlin Heidelberg New York

Springer is a part of Springer Science+Business Media

springeronline.com

© Springer-Verlag Berlin Heidelberg 2005
Printed in Germany

Typesetting: Camera-ready by author, data conversion by Scientific Publishing Services, Chennai, India
Printed on acid-free paper   SPIN: 11574781   06/3142   5 4 3 2 1 0

Preface

Peer-to-peer (P2P) computing is attracting enormous media attention, spurred by the popularity of file sharing systems such as Napster, Gnutella, and Morpheus. The peers are autonomous, or as some call them, first-class citizens. P2P networks are emerging as a new distributed computing paradigm for their potential to harness the computing power of the hosts composing the network and make their underutilized resources available to others. This possibility has generated a lot of interest in many industrial organizations that have already launched important projects.

In P2P systems, peer and Web services in the role of resources become shared and combined to enable new capabilities greater than the sum of the parts. This means that services can be developed and treated as pools of methods that can be composed dynamically. The decentralized nature of P2P computing makes it also ideal for economic environments that foster knowledge sharing and collaboration as well as cooperative and non-cooperative behaviors in sharing resources. Business models are being developed that rely on incentive mechanisms to supply contributions to the system and methods for controlling free riding. Clearly, the growth and the management of P2P networks must be regulated to ensure adequate compensation of content and/or service providers. At the same time, there is also a need to ensure equitable distribution of content and services.

Although researchers working on distributed computing, multiagent systems, databases and networks have been using similar concepts for a long time, it is only recently that papers motivated by the current P2P paradigm have started appearing in high-quality conferences and workshops. Research in agent systems in particular appears to be most relevant because, since their inception, multiagent systems have always been thought of as networks of peers.

The multiagent paradigm can thus be superimposed on the P2P architecture, where agents embody the description of the task environments, the decision-support capabilities, the collective behavior, and the interaction protocols of each peer. The emphasis in this context on decentralization, user autonomy, ease and speed of growth that gives P2P its advantages also leads to significant potential problems. Most prominent among these problems are coordination, the ability of an agent to make decisions on its own actions in the context of activities of other agents, and scalability, the value of the P2P systems lies in how well they scale along several dimensions, including complexity, heterogeneity of peers, robustness, traffic redistribution, and so on. It is important to scale up coordination strategies along multiple dimensions to enhance their tractability and viability, and thereby to widen the application domains. These two problems are common to many large-scale applications. Without coordination, agents may be wasting their efforts, squandering resources and failing to achieve their objectives in situations requiring collective effort.

This workshop brought together researchers working on agent systems and P2P computing with the intention of strengthening this connection. These objectives are accomplished by bringing together researchers and contributions from these two disciplines but also from more traditional areas such as distributed systems, networks and databases.

We sought high-quality and original contributions on the general theme of "Agents and P2P Computing". The following is a non-exhaustive list of topics of special interest:

- intelligent agent techniques for P2P computing
- P2P computing techniques for multi-agent systems
- the Semantic Web and semantic coordination mechanisms for P2P systems
- scalability, coordination, robustness and adaptability in P2P systems
- self-organization and emergent behavior in P2P systems
- e-commerce and P2P computing
- participation and contract incentive mechanisms in P2P systems
- computational models of trust and reputation
- community of interest building and regulation, and behavioral norms
- intellectual property rights and legal issues in P2P systems
- P2P architectures
- scalable data structures for P2P systems
- services in P2P systems (service definition languages, service discovery, filtering and composition etc.)
- knowledge discovery and P2P data mining agents
- P2P-oriented information systems
- information ecosystems and P2P systems
- security considerations in P2P networks
- ad-hoc networks and pervasive computing based on P2P architectures and wireless communication devices
- grid computing solutions based on agents and P2P paradigms

This volume is the postproceedings of AP2PC 2004, the 3rd International Workshop on Agents and P2P Computing,[1] held in New York City on July 19, 2004 in the context of the 3rd International Joint Conference on Autonomous Agents and Multi-agent Systems (AAMAS 2004).

These proceedings contain papers presented at AP2PC 2004 fully revised according to reviewers' comments and discussions at the workshop, plus three invited papers related to the invited talk and the panel. The volume is organized according to the following sessions held at the workshop:

- P2P networks and search performance
- emergent communities and social behaviors
- semantic integration
- mobile P2P systems

---

[1] http://p2p.ingce.unibo.it/

- adaptive systems
- agent-based resource discovery
- trust and reputation

We would like to thank the invited speaker Hector Garcia-Molina, full professor and chairman in the Department of Computer Science at Stanford University, for his talk on semantic overlay networks for P2P systems.

We would also like to thank Munindar P. Singh, full professor in the Department of Computer Science at North Carolina State University, for chairing the panel with the theme "Conducting Business via P2P and Emerging Mission-Critical Applications". We express our deepest appreciation for the workshop participants (more than 40 people) for their lively discussions, in particular for the invited panelists: Sonia Bergamaschi, Hector Garcia-Molina, Sandip Sen and Steven Willmott.

After distributing the call for papers for the workshop, we received 34 papers. All submissions were reviewed for scope and quality, 12 were accepted as full papers and 8 as short contributions. We would like to thank the authors for their submissions and the members of the Program Committee for reviewing the papers under time pressure and for their support for the workshop. Finally, we would like to acknowledge the Steering Committee for its guidance and encouragement.

This workshop followed the successful second edition, which was held in conjunction with AAMAS in Melbourne, Australia in 2003. In recognition of the interdisciplinary nature of P2P computing, a sister event called the Second International Workshop on Databases, Information Systems, and P2P Computing[2] was held in Toronto in September 2004 in conjunction with the International Conference on Very Large Data Bases (VLDB).

Fall 2004   Gianluca Moro (University of Bologna)
          Sonia Bergamaschi (University of Modena and Reggio-Emilia)
                                              Karl Aberer (EPFL)

---

[2] http://dbisp2p.ingce.unibo.it/

Executive Committee

Organizers

Program Co-chairs Gianluca Moro
Dept. of Electronics, Computer Science and Systems,
University of Bologna, Italy

Sonia Bergamaschi
Dept. of Science Engineering,
University of Modena and Reggio-Emilia, Italy

Karl Aberer
École Polytechnique Fédérale de Lausanne (EPFL)
Switzerland

Panel Chair Munindar P. Singh
Dept. of Computer Science,
North Carolina State University, USA

Steering Committee

The Steering Committee consists of the above plus the following people:

Manolis Koubarakis, Dept. of Electronic and Computer Engineering,
Technical University of Crete, Greece

Paul Marrow, Intelligent Systems Laboratory,
BTexact Technologies, UK

Aris M. Ouksel, Dept. of Information and Decision Sciences,
University of Illinois at Chicago, USA

Claudio Sartori,
IEIIT-BO-CNR, University of Bologna, Italy

Webmaster of Sam Joseph
Review System Laboratory for Interactive Learning Technology (LILT),
University of Hawaii, USA

Program Committee

Karl Aberer, EPFL, Lausanne, Switzerland
Sonia Bergamaschi, University of Modena and Reggio-Emilia, Italy
Jon Bing, University of Oslo, Norway
M. Brian Blake, Georgetown University, USA
Rajkumar Buyya, University of Melbourne, Australia
Ooi Beng Chin, National University of Singapore, Singapore
Paolo Ciancarini, University of Bologna, Italy
Costas Courcoubetis, Athens University of Economics and Business, Greece
Yogesh Deshpande, University of Western Sydney, Australia
Asuman Dogac, Middle East Technical University, Turkey
Boi V. Faltings, EPFL, Lausanne, Switzerland
Maria Gini, University of Minnesota, USA
Dina Q. Goldin, University of Connecticut, USA
Chihab Hanachi, University of Toulouse, France
Mark Klein, Massachusetts Institute of Technology, USA
Matthias Klusch, DFKI, Saarbrücken, Germany
Yannis Labrou, PowerMarket Inc., USA
Tan Kian Lee, National University of Singapore, Singapore
Zakaria Maamar, Zayed University, UAE
Dejan Milojicic, Hewlett-Packard Labs, USA
Alberto Montresor, University of Bologna, Italy
Luc Moreau, University of Southampton, UK
Jean-Henry Morin, University of Geneva, Switzerland
John Mylopoulos, University of Toronto, Canada
Andrea Omicini, University of Bologna, Italy
Maria Orlowska, University of Queensland, Australia
Aris. M. Ouksel, University of Illinois at Chicago, USA
Mike Papazoglou, Tilburg University, Netherlands
Terry R. Payne, University of Southampton, UK
Paolo Petta, Austrian Research Institute for AI, Austria
Jeremy Pitt, Imperial College London, UK
Dimitris Plexousakis, Institute of Computer Science, FORTH, Greece
Martin Purvis, University of Otago, New Zealand
Omer F. Rana, Cardiff University, UK
Katia Sycara, Robotics Institute, Carnegie Mellon University, USA
Douglas S. Reeves, North Carolina State University, USA
Thomas Risse, Fraunhofer IPSI, Darmstadt, Germany
Pierangela Samarati, University of Milan, Italy
Giovanni Sartor, CIRSFID, University of Bologna, Italy
Christophe Silbertin-Blanc, University of Toulouse, France
Maarten van Steen, Vrije Universiteit, Netherlands
Markus Stumptner, University of South Australia, Australia
Peter Triantafillou, Technical University of Crete, Greece
Anand Tripathi, University of Minnesota, USA

Vijay K. Vaishnavi, Georgia State University, USA
Francisco Valverde-Albacete, Universidad Carlos III de Madrid, Spain
Maurizio Vincini, University of Modena and Reggio-Emilia, Italy
Fang Wang, BTexact Technologies, UK
Gerhard Weiss, Technische Universität München, Germany
Bin Yu, North Carolina State University, USA
Franco Zambonelli, University of Modena and Reggio-Emilia, Italy

Additional Reviewers and Helpers

We would like to thank the following additional reviewers for their valuable help:

- Luca Caviglione
- Julio Cesar Hernandez Castro
- Lican Huang
- Samuel Joseph
- Stefano Morini
- Wolfgang Mueller
- Ben Strulo
- Dimitrios Tsoumakos
- Alessandra Villecco
- Emily Weitzanböck

Finally thanks also to Jonathan Gelati for his technical help in putting together the electronic version of this volume.

Preceding Editions of AP2PC

Here are the references to the preceding editions of AP2PC, including the volumes of revised and invited papers:

- AP2PC 2002 was held in Bologna, Italy, July 15th, 2002. The website can be found at http://p2p.ingce.unibo.it/2002/. The proceedings were published by Springer as LNCS Vol. 2530 and are available online here: http://www. springerlink.com/link.asp?id=6qr2pb576my5
- AP2PC 2003 was held in Melbourne, Australia, July 14th, 2003. The website can be found at http://p2p.ingce.unibo.it/2003/. The proceedings were published by Springer as LNCS Vol. 2872.

Table of Contents

Invited Talk

Semantic Overlay Networks for P2P Systems
   Arturo Crespo, Hector Garcia-Molina............................ 1

Peer-to-Peer Network and Search Performance

Unstructured Peer-to-Peer Networks: Topological Properties and
Search Performance
   George H.L. Fletcher, Hardik A. Sheth, Katy Börner 14

Distributed Hash Queues: Architecture and Design
   Chad Yoshikawa, Brent Chun, Amin Vahdat..................... 28

DiST: A Scalable, Efficient P2P Lookup Protocol
   Savitha Krishnamoorthy, Karthikeyan Vaidyanathan,
   Mario Lauria.. 40

A Policy for Electing Super-Nodes in Unstructured P2P Networks
   Georgios Pitsilis, Panayiotis Periorellis, Lindsay Marshall 54

Emergent Communities and Social Behaviours

ACP2P: Agent Community Based Peer-to-Peer Information Retrieval
   Tsunenori Mine, Daisuke Matsuno, Akihiro Kogo,
   Makoto Amamiya 62

Emergent Structures of Social Exchange in Socio-cognitive Grids
   Daniel Ramirez-Cano, Jeremy Pitt 74

Permission and Authorization in Policies for Virtual Communities of
Agents
   Guido Boella, Leendert van der Torre 86

On Exploiting Agent Technology in the Design of Peer-to-Peer
Applications
   Steven Willmott, Josep M. Pujol, Ulises Cortés.................. 98

Semantic Integration

Peer-to-Peer Semantic Integration of XML and RDF Data Sources
Isabel F. Cruz, Huiyong Xiao, Feihong Hsu 108

The SEWASIE Multi-agent System
Sonia Bergamaschi, Pablo R. Fillottrani, Gionata Gelati 120

Mobile P2P Systems

Service Discovery on Dynamic Peer-to-Peer Networks Using Mobile
Agents
Evan A. Sultanik, William C. Regli 132

An Agent Module for a System on Mobile Devices
*Praveen Madiraju, Sushil K. Prasad, Rajshekhar Sunderraman,
Erdogan Dogdu* 144

Multi-agent System Technology for P2P Applications on Small Portable
Devices
*Martin Purvis, Noel Garside, Stephen Cranefield,
Mariusz Nowostawski, Marcos De Oliveira* 153

Adaptive Systems

Coordinator Election Using the Object Model in P2P Networks
Hirokazu Yoshinaga, Takeshi Tsuchiya, Keiichi Koyanagi 161

The Dynamics of Peer-to-Peer Tasks: An Agent-Based Perspective
Xiaolong Jin, Jiming Liu, Zhen Yang 173

Peer-to-Peer Computing in Distributed Hash Table Models Using a
Consistent Hashing Extension for Access-Intensive Keys
Arnaud Dury 185

A Practical Peer-Performance-Aware DHT
*Yan Tang, Zhengguo Hu, Yang Zhang, Lin Zhang,
Changquan Ai* 193

Agent-Based Resource Discovery

Peer-to-Peer Data Lookup for Multi-agent Systems
Michael Thomas, William Regli 201

Intelligent Agent Enabled Genetic Ant Algorithm for P2P Resource
Discovery
    Prithviraj(Raj) Dasgupta   213

Photo Agent: An Agent-Based P2P Sharing System
    Jane Yung-jen Hsu, Jih-Yin Chen, Ting-Shuang Huang,
    Chih-He Chiang, Chun-Wei Hsieh   221

Trust and Reputation

How Social Structure Improves Distributed Reputation
Systems - Three Hypotheses
    Philipp Obreiter, Stefan Fähnrich, Jens Nimis   229

Opinion Filtered Recommendation Trust Model in Peer-to-Peer
Networks
    Weihua Song, Vir V. Phoha   237

Author Index   245

Semantic Overlay Networks for P2P Systems

Arturo Crespo and Hector Garcia-Molina

Stanford University
{crespo, hector}@cs.stanford.edu

Abstract. In a peer-to-peer (P2P) system, nodes typically connect to a small set of random nodes (their neighbors), and queries are propagated along these connections. Such query flooding tends to be very expensive. We propose that node connections be influenced by content, so that for example, nodes having many "Jazz" files will connect to other similar nodes. Thus, semantically related nodes form a Semantic Overlay Network (SON). Queries are routed to the appropriate SONs, increasing the chances that matching files will be found quickly, and reducing the search load on nodes that have unrelated content. We have evaluated SONs by using an actual snapshot of music-sharing clients. Our results show that SONs can significantly improve query performance while at the same time allowing users to decide what content to put in their computers and to whom to connect.

1 Introduction

Peer-to-peer systems (P2P) have grown dramatically in recent years. They offer the potential for low cost sharing of information, autonomy, and privacy. However, query processing in current P2P systems is very inefficient and does not scale well. The inefficiency arises because most P2P systems create a random overlay network where queries are blindly forwarded from node to node. As an alternative, there have been proposals for "rigid" P2P systems that place content at nodes based on hash functions, thus making it easier to locate content later on (e.g., [1, 2]). Although such schemes provide good performance for point queries (where the search key is known exactly), they are not as effective for approximate, range, or text queries. Furthermore, in general, nodes may not be willing to accept arbitrary content nor arbitrary connections from others.

In this paper we propose Semantic Overlay Networks (SONs), a flexible network organization that improves query performance while maintaining a high degree of node autonomy. With Semantic Overlay Networks (SONs), nodes with semantically similar content are "clustered" together. To illustrate, consider Figure 1 which shows eight nodes, A to H, connected by the solid lines. When using SONs, nodes connect to other nodes that have semantically similar content. For example, nodes A, B, and C all have "Rock" songs, so they establish connections among them. Similarly, nodes C, E, and F have "Rap" songs, so they cluster close to each other. Note that we do not mandate how connections are done inside a SON. For instance, in the Rap SON node C is not required to connect directly to F. Furthermore, nodes can belong to more than one SON

G. Moro, S. Bergamaschi, and K. Aberer (Eds.): AP2PC 2004, LNAI 3601, pp. 1–13, 2005.

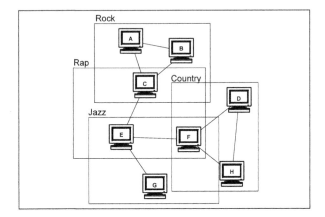

Fig. 1. Semantic Overlay Networks

(e.g., C belongs to the Rap and Rock SONs). In addition to the simple partition-ing illustrated by Figure 1, in this paper we will also explore the use of content hierarchies, where for example, the Rock SON is subdivided into "Soft Rock" and "Hard Rock."

In a SON system, queries are processed by identifying which SON (or SONs) are better suited to answer it. Then the query is sent to a node in those SONs and the query is forwarded only to the other members of that SON. In this way, a query for Rock songs will go directly to the nodes that have Rock content (which are likely to have answers for it), reducing the time that it takes to answer the query. Almost as important, nodes outside the Rock SON (and therefore unlikely to have answers) are not bothered with that query, freeing resources that can be used to improve the performance of other queries.

There are many challenges when building SONs. First, we need to be able to classify queries and nodes (what does "contain rock songs" means?). We need to decide the level of granularity for the classification (e.g., just rock songs versus soft, pop, and metal rock) as too little granularity will not generate enough locality, while too much would increase maintenance costs. We need to decide when a node should join a SON (if a node has just a couple of documents on "rock," do we need to place it in the same SON as a node that has hundreds of "rock" documents?). Finally, we need to choose which SONs to use when answering a query.

Many of our questions can only be answered empirically by studying real P2P content and how well it can be organized into SONs. For our empirical evaluation we have chosen music-sharing systems. These systems are of interest not only because they are the biggest P2P application ever deployed, but also because music semantics are rich enough to allow different classification hierarchies. In addition there is a significant amount of data available that allows us to perform realistic evaluations. While our experimental results in this paper are particular to this important application, we have no reason to believe they would not apply in other applications with good classification hierarchies.

Also note that due to space limitations, in this paper we do not present the full results of our work. A more detailed and formal description of our approach, as well as additional experimental results, can be found in the extended version of this paper [3].

2   Related Work

The idea of placing data in nodes close to where relevant queries originate was used in early distributed database systems [4]. However, the algorithms used for distributed databases are based on two fundamental assumptions that are not applicable to P2P systems: that there are a small number of stables nodes, and that the designer has total control over the data. There are a number of P2P research systems (CAN [2], CHORD [1], Oceanstore [5], Pastry [6], and Tapestry [7]) that are designed so documents can be found with a very small number of messages. However, all these techniques either mandate a specific network structure or assume total control over the location of the data. Although these techniques may be appropriate in some application, the lack of node autonomy has prevented their use in wide-scale P2P systems. There is a large corpus of work on document clustering using hierarchical systems (see [8] for a survey). However, most clustering algorithms assume that documents are part of a controlled collection located at a central database. Clustering algorithms for decentralized environments have also been studied in the context of the web. However, these techniques depend on crawling the data into a centralized site and then using clustering techniques to either make web search results more accurate (as in SONIA [9]) or easier to understand. A more decentralized approach has been taken by Edutella [10].

3   Semantic Overlay Networks

In this section we introduce informally the concept of Semantic Overlay Networks (SONs) (see [3] for a formal definition). In a P2P system, the links between the nodes typically form a single overlay network. In this paper we advocate the creation of multiple overlay networks to improve search performance. We do *not* focus on on how queries are routed within an overlay network (see Section 2 for a brief overview of current solutions to the intra-overlay network routing problem). Therefore, we will ignore the link structure within an overlay network and we will represent an overlay network just by the set of nodes in it.

Requests for documents are made by issuing a query q and some additional system-dependent information (such as the horizon of the query). A query is also system dependent and it can be as simple as a document identifier, or keywords, or even a complex SQL query. In this paper we assume that queries are partial, so the request includes a minimum number of results that need to be returned.

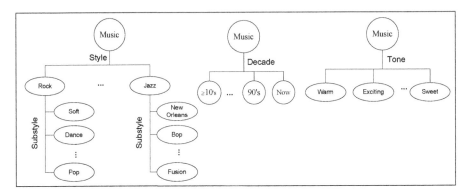

Fig. 2. Classification Hierarchies

3.1   Classification Hierarchies

Our objective is to define a set of overlay networks in such a way that, when given a request, we can select a small number of overlay networks whose nodes have a "high" number of hits. The benefit of this strategy is two fold. First, the nodes to which the request is sent will have many matches, so the request is answered faster; and second, but not less important, the nodes that have few results for this query will not receive it, avoiding wasting resources on that request (and allowing other requests to be processed faster).

We propose using a classification hierarchy as the basis of the formation of the overlay networks. For example, in Figure 2, we show 3 possible classification hierarchies for music documents. In the first one, music documents are classified according to their style (rock, jazz, etc.) and their substyle (soft, dance, etc.); in the second one, they are classified by decade; and in the third one, they are classified by tone (warm, exciting, etc.).

Each document and query is classified into one or more *leaf* concepts in the hierarchy. However, in practice, classification procedures may be *imprecise* as they may not able to determine exactly to which concept a query or document belongs. In this case, imprecise classification functions may return non-leaf concepts, meaning that document or query belongs to one or more descendant of the non-leaf concept, but the classifier cannot determine which one. For example, when using the leftmost classification hierarchy of Figure 2, a "Pop" document may be classified as "Rock" if the classifier cannot determine to which substyle ("Pop," "Dance," or "Soft") the document actually belongs. Classifiers may also make *mistakes* by returning the wrong concept for a query or document. We call the set of documents that classify into the same concept the "bucket" of that concept.

In a P2P system, documents are actually kept by nodes. Therefore, we need to classify nodes, rather than documents. We call a group of semantically related nodes a Semantic Overlay Network (SON), and we associate each SON with a concept in the classification hierarchy. We call a SON associated with concept c, the SON for c or SON_c. For example, in the leftmost hierarchy in Figure 2 (if we

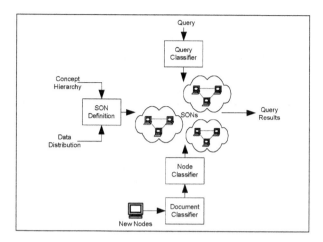

Fig. 3. Generating Semantic Overlay Networks

assume that only the its only concepts are the ones shown), we will define at 9
SONs: 6 associated with the leaf nodes (soft, dance, pop, New Orleans, etc.), one
associated with rock, another associate with jazz, and a final one associate with
music. To completely define a SON, we need to explain how nodes are assigned
to SONs and how we decide which SONs to use to answer a query.

A node decides which SONs to join based on the classification of its documents. There are many strategies for node placement. For example, we may
place a node in SON_c if it has any document classified in c. This strategy is very
conservative as it will place a node in SON_c if just one document classifies as c.
A less conservative strategy will place a node in SON_c if a "significant" number of document classifies as c. Such less-conservative strategy has two effects:
it reduces the number of nodes in a SON and it reduces the number of SONs
to which a node belongs. The first of these effects increases the advantages of
SONs as fewer nodes need to be queried. The second effect reduces the cost of
SONs as the greater the number of SONs to which a node belongs, the greater
the the node overhead for handling many different connections. However, a less
conservative strategy may prevent us from finding all documents that match a
query. In the extended version of this paper, we study different strategies for
assignment of nodes to SONs.

After assigning nodes to SON, we may make adjustments to the SONs based
on the actual data distributions in the nodes. For example, if we observe that a
SON contains only a very small number of nodes, we may want to consolidate
that SON with a sibling or its parent in order to reduce overhead.

To summarize, the process of building and using SONs is depicted in Figure 3. First, we evaluate potential classification hierarchies using the actual data
distributions in the nodes (or a sample of them) and find a good hierarchy. This
hierarchy will be stored by all (or some) of the nodes in the system and it is
used to define the SONs. A node joining the system, first floods the network
with requests for the hierarchy in a Gnutella fashion (we do not address se-

curity problems in this paper, but inconsistent hierarchies may be detected by obtaining the hierarchy from multiple sources and using a majority rule). Then, the node runs a document classifier based on the hierarchy obtained on all its documents. Then, a node classifier assigns the node to specific SONs (by, for example, using the conservative strategy described in this section). The node joins each SON by finding nodes that belong to those SONs. This can be done again in a Gnutella fashion (flooding the network until nodes in that SON are found) or by using a central directory. When the node issues a query, first it classifies it and sends it to the appropriate SONs (nodes in those SONs can be found in a similar fashion as when the node connected to its SON). After the query is sent to the appropriate SONs, nodes within the SON find matches by using some propagation mechanism (such as Gnutella flooding or super peers).

In the next sections, we will study the challenges and present solutions for building a P2P system using Semantic Overlay networks. We will evaluate our solutions by simulating a music-sharing system based on real data from Napster [11] and OpenNap [12]. Specifically, in this paper we will address the following challenges:

- Classifying queries and documents (Section 4): Imprecise classifiers can map too many documents and queries to higher levels of the hierarchy, making searches more expensive. What are the options for building classifiers? Are they precise enough for our needs? What is the impact of classification errors?
- Searching SONs (Section 6): How do we search SONs? Is it worth having Semantic Overlay Networks? Is the search performance of a SON-based system better than a single-overlay network system such as Gnutella?
- SON membership When should a node join a SON? What is the cost of joining a SON? Can we reduce the number of SONs that a node needs to belong to (while being able to find most results)?

4   Classifying Queries and Documents

In this section we describe how documents and queries are classified. Although the problem of classifying documents and the problem of classifying queries are very similar, the *requirements* for the document and query classifiers can be very different. Specifically, it is reasonable to expect that nodes will join a relatively stable P2P network at a low rate (a few per minute); while we could expect a much higher query rate (hundreds or even more per second). Additionally, node classification is more bursty as when a node joins the network it may have hundreds of documents to be classified; on the other hand, queries will likely to arrive at a more regular rate. Under these conditions, the document classifier can use a very precise (but time consuming) algorithm that can process in batch a large number of documents; while, the query classifier must be implemented by a fast algorithm that may have to be imprecise.

The classification of documents and queries can be done automatically, manually, or by a hybrid processes. Examples of automatic classifiers include text matching [13], Bayesian networks [14], and clustering algorithms [15]. These automatic techniques have been extensively studied and they are beyond the scope of this paper. Manual classification may be achieved by requiring users to tag each query with the style or substyle of the intended results. If the user does not know the substyle or style of the potential results, he can always select the root of the hierarchy so all nodes are queried. Finally, hybrid classifiers aid the manual classification with databases as we will see shortly in our experiments.

4.1   Evaluating Our Document Classifier

Documents were classified by probing the database of All Music Guide at allmusic.com [16]. In this database songs and artists are classified using a hierarchy of style/substyle concepts equivalent to the leftmost classification hierarchy of Figure 2. Recall that for each Napster node used in our evaluation we had a list of filenames with the format "directory/author-song title.mp3." As a first step, the document classifier extracted the author and the song title for the file. The classifier then probed the database with that author and song and obtained a list of possible song matches. Finally, the classifier selected the highest rank song and found its style and substyles. If there were not matches in the database, the classifier assigned "unknown" to the style and substyle of the file.

There were many sources of errors when using our document classifier. First, the format of the files may not follow the expected standard, so the extraction of the author and song title may return erroneous values. Second, we assumed that all files were music (but Napster could be, and was actually used, to share other kind of files). Third, users made misspellings in the name of artist and/or song (to reduce the effect of misspellings, we used a phonetic search in the All Music database, so some common misspellings did not affect the classification). Finally, the All Music database is not complete, which is especially true in the case of classical music.

To evaluate the document classifier, we measured the number of incorrect classifications. We selected 200 random filenames and manually found the substyles to which they belong (occasionally using the All Music database and

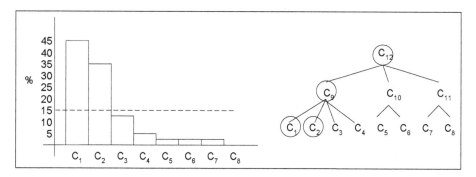

Fig. 4. Choosing SONs to join

Google as an aid to find the substyles of obscure pieces). We then compared the manual classification with the one obtained from our document classifier. We considered a classification to be incorrect for a given document if the document classifier returned one or more substyles to which the document should not belong. Note that an "unknown" classification from our classifier, although very imprecise, is not incorrect as it would correspond to the root node of the classification hierarchy. In our evaluation, we found that 25% of the files were classified incorrectly.

However, a node can still be correctly classified even if some of its documents are misclassified. (If a node is properly classified, it will be possible to find the misclassified documents later on.) To evaluate the true effect of document misclassification, we selected 20 random nodes, we classified all their documents, and assigned the nodes to all the substyles of their respective documents. We considered a classification to be incorrect for a given *node* if the node was not assigned to one or more substyles to which the node should belong. In our evaluation, we found that only 4% of the nodes were classified incorrectly. This result shows that errors when classifying documents tend to cancel each other within a node. Specifically, even if we fail to classify a document as, for example, "Pop" it is likely that there will be some other "Pop" document in the node that will be classified correctly so the node will still be in the "Pop" SON.

4.2   Evaluating Our Query Classifier

For our experiments, queries were classified by hand by the authors of this paper. Queries were either classified in one or more substyles, a single style, or as "music"(the root of the hierarchy). In our experiments we used queries obtained from traces of actual queries sent to an OpenNap server run at Stanford [17]. Thus, by manually classifying queries, we are "guessing" what the users would have selected from say a drop-down menu as they submitted their queries.

Unfortunately, we cannot evaluate the correctness of the query classification method (we, of course, consider our classification of all queries to be correct). Nevertheless, we can study how precise our manual classification was (i.e., how many times queries were classified into a substyle, a style, or at the root of the classification hierarchy). We selected a trace of 50 *distinct* queries (the original query trace contained many duplicates which the authors of [17] believed were the result of cycles in the OpenNap overlay network) and then manually classified those queries. The result was that 8% of the queries were classified at the root of the hierarchy, 78% were classified a the style level of the hierarchy and 14% at the substyle level. In the extended version of this paper we show that the distribution of queries over hierarchy levels impacts the overall system performance, as more precisely classified queries can be executed more efficiently.

5   Nodes and SON Membership

In Section 3 we presented a conservative strategy for nodes to decide which SONs to join. Basically, under this strategy, nodes join all the SONs associated with a concept for which they have a document. This strategy guarantees that we will be able to find all the results, but it may increase both the number of nodes in each SON and the number of connections that a node needs to maintain. A less conservative strategy, where nodes join some of all the possible SONs, can have better performance. In this section we introduce a non-conservative assignment strategy: Layered SONs.

The Layered SONs approach exploits the very common zipfian data distribution in document storage systems. (It has been shown that the number of documents in a website when ranked in order of decreasing frequency, tend to be distributed according to Zipf's Law [18].) For example, on the left side of Figure 4 we present a hypothetical histogram for a node with a zipfian data distribution (we'll explain the rest of the figure shortly). In this histogram we can observe that 45% of the documents in the node belong to category c_1, about 35% of the documents belong to category c_2, while the remaining documents belong to categories c_3 to c_8. Thus, which SONs should the node join? The conservative strategy mandates that the node need to join SON_{c_1} through SON_{c_8}. However, if we assume that queries are uniform over all the documents in a category, it is clear that the node will have a higher probability of answering queries in SON_{c_1} and SON_{c_2} than queries in the other SONs. In other words, the benefit of having the node belong to SON_{c_1} and SON_{c_2} is high, while the benefit of joining the other SONs will be very small (and even negative due to the overhead of SONs). A very simple and aggressive alternative would be to have the node join only SON_{c_1} and SON_{c_2}. However, this alternative would prevent the system from finding the documents in the node that do not belong to categories c_1 and c_2.

Nodes determine which SONs to join based on the number of documents in each category. To illustrate, consider again Figure 4. At the right of the figure we present the hierarchy of concepts that will aid a node in deciding which SONs to join. In addition, a parameter of the Layered SON approach is the minimum percentage of documents that a node should have in a category to belong to the associated SON (alternatively, we can also use an absolute number of documents instead of a percentage). In the example, we have set that number at 15%. Let us now determine which SONs the node with the histogram at the left of Figure 4 should join. First, we consider all the base categories in the hierarchy tree (c_1 to c_8). As c_1 and c_2 are above 15%, the node joins SON_{c_1} and SON_{c_2}. As all the remaining categories are all below 15%, the node does not join their SONs. We then consider the second level categories (c_9, c_{10}, and c_{11}). As the combination of the non-assigned descendants of c_9, c_3 and c_4, is higher than 15%, the node joins SON_{c_9}. However, the node does not join the SON of c_{10} as the combination of c_5 and c_6 are not above 15%. Similarly the node does not join the SONs of c_{11} as c_7 and c_8 are below the threshold. Finally, the node joins the SON associated with the root of the tree ($SON_{c_{12}}$) as there were categories (c_5, c_6, c_7 and c_8) that are not part of any assignment. This final assignment is done regardless of the 15% threshold as this ensures that all documents in the node can be found (in our example, if we do not join $SON_{c_{12}}$ we will not be able to find the documents in the SONs of c_5, c_6, c_7 and c_8).

The conservative assignment is equivalent to a Layered SON where the threshold for joining a SON has been set to 0%. In this case, the node will join the SONs associated with all the base concepts for which it has one or more documents.

6   Searching SONs

In this section, we explore the problem of how to choose among a set of SONs when using Layered SONs. (We discussed in Section 3.1 the mechanisms used by nodes to actually send the queries to those SONs.)

6.1   Searching with Layered SONs

Searches in Layered SONs are done by first classifying the query. Then, the query is sent to the SON (or SONs) associated with the base concept (or concepts) of the query classification. Finally, the query is progressively sent higher up in the hierarchy until enough results are found. In case more than one concept is returned by the classifier, we do a sequential search in all the concepts returned before going higher up in the hierarchy. For example, when looking for a "Soft Rock" file we start with the nodes in the "Soft Rock" SON. If not enough results are found (recall that partial queries have a target number of results), we send the query to the "Rock" SON. Finally, if we still have not found enough results, we send the query to the "Music" SON. There are multiple approaches when searching with Layered SONs. In this paper we are concentrating on a single serial one (as our objective is to minimize number of messages). However, there are other approaches such as searching more than one SON in parallel (by asking each one for some fraction of the target results) which may result in higher number of messages, but will start producing results faster.

This search algorithm does not guarantee that all documents will be found if there are classification mistakes for documents. Not finding all documents may or may not be a problem depending on the P2P system, but in general, if we need to find all documents for a query (in the presence of classification mistakes), our only option is an exhaustive search among all nodes in the network. However, we will see that with our document classifier (which has an per-document classification mistake probability of 25%), we can find more than 95% of the documents that match a query. In addition, this search algorithm may result in duplicate results. Specifically, duplication can happen when a node belongs, at the same time, to a SON associated with a substyle and to the SON associated with the parent style of that substyle. In this case, a query that is sent to both SONs will search the node twice and thus it will find duplicate results.

6.2   Experiments

We will now consider two possible SON configurations and evaluate their performance against a Gnutella-like system. As before, we used the crawl of 1800 Napster nodes made at the University of Washington, which were classified using

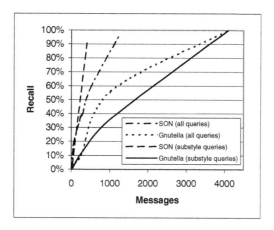

Fig. 5. Average number of Messages for a Query Trace

the All Music database. We assumed that the nodes in the network (both inside SONs and in the Gnutella network) were connected via an acyclic graph and that on average each node was connected to four other nodes. Although the assumption of an acyclic graph is not realistic, we are considering acyclic networks as the effect of cycles is independent of the creation of SONs. Cycles affect a P2P system by creating repeated messages containing queries that the receiving nodes have already seen. Therefore, an analysis of an acyclic P2P network gives us a lower estimate of the number of messages generated.

For this experiment we used 50 different random queries obtained from traces of actual queries sent to an OpenNap server run at Stanford [17]. These queries were classified by hand. Queries classified at the substyle level were sent sequentially to the corresponding SON (or SONs), and then to the style-level SON. Queries classified at the style level, were first sent sequentially to all substyles of that style, and then to the style level. Queries classified at the root of the hierarchy were sent to all nodes. We measure the level of recall averaged for all 50 queries versus the number of messages sent in the system. The graphs were obtained by running 50 simulations over randomly generated network topologies.

In Figure 5, we show the result of this experiment. The figure shows the number of messages sent versus the level of recall. Layered SONs were able to obtain the same level of matches with significantly fewer messages than the Gnutella-like system. However, Layered SONs do not achieve recall levels of 100% in general (average maximum recall was 93%) due to mistakes in the classification of nodes.

The results of Figure 5 show the average performance for all query types (dotted line). However, if a user is able to precisely classify his query, he will get significantly better performance. To illustrate this point, Figure 5 also shows with a dashed the number of messages sent versus the level of recall for queries classified at the substyle level (the lowest level of the hierarchy). In this case, we obtain a significant improvement versus Gnutella. For example, to obtain a recall level of 50%, Layered SONs required only 461 messages, while Gnutella needed 1731 messages, a reduction of 375% in the number of messages. Moreover, even at high recall levels, Layered SONs were able to reach a recall level of 92% with about 1/5 of the messages that Gnutella required.

The shape of the curve for the message performance of Gnutella is slightly different for all queries and for queries classified at the substyle level. The reason for this difference is very subtle. The authors of this paper were only able to classify very precisely (i.e. to the substyle level) queries for songs that are very well known. Due to their popularity, there are many copies of these songs throughout the network. Therefore, a Gnutella search approach will have a high probability of finding a match in many of the nodes visited, making the flooding of the network less of a problem than with more rare songs. Nevertheless, even in this case, Layered SONs performed much better than Gnutella.

7   Conclusion

We studied how to improve the efficiency of a peer-to-peer system by clustering nodes with similar content in Semantic Overlay Networks (SONs). We showed how SONs can efficiently process queries while preserving a high degree of node autonomy. We introduced Layered SONs, an approach that improves query performance even more at a cost of a slight reduction in the maximum achievable recall level. From our experiments we conclude that SONs offer significant improvements versus random overlay networks, while keeping costs low. We believe that SONs, and in particular Layered SONs, can help improve the search performance of current and future P2P systems where data is naturally clustered.

References

1. Stoica, I., Morris, R., Karger, D., Kaashoek, M.F., Balakrishnan, H.: Chord: A scalable peer-to-peer lookup service for internet applications. In: Proc. ACM SIG-COMM. (2001)
2. Ratnasamy, S., Francis, P., Handley, M., Karp, R., Shenker, S.: A scalable content-addressable network. In: ACM SIGCOMM. (2001)
3. Crespo, A., Garcia-Molina, H.: Semantic overlay networks for p2p systems. Technical report, Stanford University (2003) At http://dbpubs.stanford.edu/pub/2003-75.
4. Kossman, D.: The state of the art in distributed queyr processing. ACM Computing Survey (2000)
5. Kubiatowicz, J., Bindel, D., Chen, Y., Czerwinski, S., Eaton, P., Geels, D., Gummadi, R., Rhea, S., Weatherspoon, H., Weimer, W., Wells, C., Zhao, B.: Oceanstore: An architecture for global-scale persistent storage. In: ASPLOS. (2000)
6. Rowstron, A., Druschel, P.: Pastry: Scalable, distributed object location and routing for large-scale peer-to-peer systems. In: Middleware. (2001)
7. Zhao, B., Kubiatowicz, J., Joseph, A.: Tapestry: An infrastructure for fault-tolerant wide-area location and routing. Technical report, U. C. Berkeley (2001)
8. Manning, C., Schutze, H.: Foundations of statistical natural language processing. The MIT Press (1999)

 9. Sahami, M., Baldonado, S.Y.M.: Sonia: A service for organizing networked information autonomously. In: Proceedings of the Third ACM Conference on Digital Libraries. (1998)
10. Nejdl, W., Siberski, W., Wolpers, M., Schmitz, C.: (Routing and clustering in schema-based super peer networks)
11. WWW: http://www.napster.com: (Napster)
12. WWW: http://opennap.sourceforge.net: (OpenNap)
13. R. Baeza-Yates, B.R.N.: Modern Information Retrieval. Addison Wesley (1999)
14. Rich, E., Knight, K.: Artificial Intelligence. McGraw-Hill Inc. (1991)
15. Witten, I., Frank, E.: Data Mining. Morgan Kaufmann Publishers (1999)
16. WWW: http://www.allmusic.com: (All Music Guide)
17. Yang, B., Garcia-Molina, H.: Comparing hybrid peer-to-peer systems. In: Proceedings of the Tweenty-first International Conference on Very Large Databases (VLDB'01). (2001)
18. Korfhage, R.: Information storage and retrieval. Wiley Computer Publishing (1997)

Unstructured Peer-to-Peer Networks: Topological Properties and Search Performance

George H.L. Fletcher*, Hardik A. Sheth**, and Katy Börner***

*Computer Science Department
** School of Informatics
*** School of Library and Information Science,
Indiana University, Bloomington, USA
{gefletch, hsheth, katy}@indiana.edu

Abstract. Performing efficient decentralized search is a fundamental problem in Peer-to-Peer (P2P) systems. There has been a significant amount of research recently on developing robust self-organizing P2P topologies that support efficient search. In this paper we discuss four structured and unstructured P2P models (CAN, Chord, PRU, and Hypergrid) and three characteristic search algorithms (BFS, k-Random Walk, and GAPS) for unstructured networks. We report on the results of simulations of these networks and provide measurements of search performance, focusing on search in unstructured networks. We find that the proposed models produce small-world networks, and yet none exhibit power-law degree distributions. Our simulations also suggest that random graphs support decentralized search more effectively than the proposed unstructured P2P models. We also find that on these topologies, the basic breadth-first search algorithm and its simple variants have the lowest search cost.

1 Introduction

Peer-to-Peer (P2P) networks have sparked a great deal of interdisciplinary excitement and research in recent years [17]. This work heralds a fruitful perspective on P2P systems vis-á-vis open multi-agent-systems (MAS)[1] [14]. A central issue for both P2P networks and MAS is the problem of decentralized search; an effective search facility that uses only local information is essential for their scalability and, ultimately, their success. Initial work on this issue suggests that there is a strong relationship between network topology and search algorithms; several deployed P2P networks [3,10,11,18] and MAS [2] have been shown to exhibit

---

[1] In an *open* MAS, agents do not have complete global knowledge of system membership.

G. Moro, S. Bergamaschi, and K. Aberer (Eds.): AP2PC 2004, LNAI 3601, pp. 14–27, 2005.

power-law degree distributions[2] and small-world properties.[3] In a small-world network, there is a short path between any two nodes. This knowledge, however does not give much leverage during search for paths in small-world systems because there are no local clues for making good choices. What is the best we can do for decentralized search in a small-world? There has been little comparative analysis of unstructured P2P models and search algorithms. Such validation and comparison of models and algorithms is the first step in answering this question.

The approach we have taken to explore this issue is to model the network topologies of two typical unstructured P2P models developed in the P2P community (PRU [19] and Hypergrid [21]) in simple graph-theoretic terms and build simulations of these networks to measure topological properties and search performance. As a comparison, we performed the same analyses on a random graph [3,18] and two structured P2P models (CAN [20] and Chord [24]). We show through these simulations that unstructured P2P networks have exactly the properties and problems of small-world topologies; the networks have low diameter but no means of directing search efficiently. Interestingly, these simulations also show that none of the models considered generate power-law degree distributions. This turns out to be desirable in an engineered system; although power-law networks support efficient decentralized search [1], they are fragile in the face of attack [3] and can unfairly distribute network traffic during search [21]. The reason for these weaknesses lies in the degree distribution; such networks have a few nodes of very high degree that serve effectively as local "hubs."

1.1   P2P Concepts and Related Work

There are two broad categories of P2P systems: hybrid and pure [17]. Hybrid systems are characterized by some form of centralized control such as a name look-up service [17] or a middle agent [8]. Pure systems strive for self-organization and total decentralization of computation; these systems are the focus of the work presented in this paper.

Pure P2P networks can be classified by the manner in which decentralization is realized. In *structured* systems [20,24], placement of system resources at nodes is strictly controlled and network evolution, consequently, incurs extra overhead. Ideally, one would strive to minimize system constraints and costly datastructures when designing a P2P model. *Unstructured* systems are characterized by a complete lack of constraints on resource distribution and minimal network growth policies. These systems focus on growing a network with the desirable low diameter of small world systems using only limited local information.

Early work on search methods for small world networks was done by Walsh [27] and Kleinberg [13] and on decentralized search in scale-free networks by

---

[2] The degree distribution of nodes in a graph follows a power-law if the probability $P(k)$ that a randomly chosen node has k edges is $P(k) \propto k^{-\tau}$, for τ a constant skew factor [3,18].

[3] A small-world network is characterized by low diameter and high clustering coefficient, relative to a random graph of equivalent size [28]. We will define these properties in full below.

Adamic et al. [1]. An early study of unstructured P2P network search performance was done by Lv et al. [15], comparing search performance on generic power-law, random, and Gnutella networks.[4] More recently, several groups have continued to study search performance with a focus on comparing power-law and random topologies with deployed P2P systems such as Gnutella [5,25,29]. Initial studies on search in open MAS have also focused on generic topologies [9,23]. Several projects have investigated the topological characteristics of the Internet [10] and implementations of P2P filesharing networks [11]. What has been missing in all of this work is a general comparative study of proposed unstructured P2P models, their topologies, and performance of search algorithms. This paper is an initial step in filling this gap in our understanding of decentralized search in unstructured P2P networks and open MAS.

2   P2P Models

In this section, we briefly introduce the P2P models under discussion. To facilitate comparison, we consider network topologies using a uniform graph-theoretic framework. We view peers as nodes in an undirected graph of size $\mathcal{M}$ where edges indicate connections between peers in the network. Each node N in the graph has, as an attribute, a routing table $\mathcal{T}_N = [e_1 : w_1, \ldots, e_k : w_k]$ that associates a weight w_i to each edge e_i ($1 \leqslant i \leqslant k$) incident on N. This represents the connections of node N to k neighbors in the graph. Unless otherwise stated, all weight values are equal in the graph.

2.1   Structured Models

As mentioned above, structured models enforce strict constraints on network evolution and resource placement. These constraints limit network robustness and node autonomy. Structured P2P models are good for building systems where controlled resource placement is a high priority, such as distributed file storage. However, they are not good models for systems with highly dynamic membership. The main advantage of these models is that the added constraints result in sublinear search mechanisms; each of these models has an associated native search mechanism that takes advantage of the added structure [20,24].

CAN. The Content Addressable Network (CAN), proposed by Ratnasamy et al. [20], is a framework for structured P2P systems based on a virtual d-dimensional Cartesian coordinate space on a d-torus. Nodes in a CAN graph have as an attribute the coordinates of a subspace of this space that are used in adding nodes and edges to the graph. Initially, the graph consists of one node and no edges. This initial node is assigned the entire virtual space. As nodes are added to the graph, they are assigned a subspace in the virtual space from a uniform distribution. The system self-organizes to adjust to a new node by

---

[4] http://www.gnutella.com

 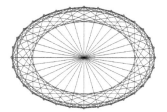

Fig. 1. 32 Node CAN and Chord Networks

adding edges from the new node to adjacent nodes in the space. A visualization of a 32 node CAN graph is given in Figure 1 on the left.[5]

Chord. Chord, proposed by Stoica et al. [24], is another self-organizing structured P2P system model. Nodes in a Chord graph have, as an additional attribute, a coordinate in a 1−dimensional virtual space (called a *ring*). When a new node N is added to the graph, the routing table attributes of the nodes adjacent to N on the ring are used to add edges between N and k other nodes distributed in the space. A visualization of a 32 node Chord graph is given in Figure 1 on the right.

2.2   Unstructured Models

Unstructured models strive for complete decentralization of decision making and computation. They require only local maintenance procedures and are topologically robust in the face of system evolution. These models are good for building highly dynamic systems where anonymity and minimal administrative overhead are prized.

Random Graph. We utilize the *Erdös-Rényi random graph* as a baseline model for comparison with unstructured networks [3,18]. There is one parameter in building a system with this topology: connection probability p. To build a random network based on this model, the graph initially has no edges. Then for each possible undirected edge between two distinct nodes in the graph, an edge is added with probability p.

PRU. The PRU (Pandurangan-Raghavan-Upfal) model for unstructured systems, proposed by Pandurangan et al. [19], is based on a simple network growth policy that ensures low graph diameter. In these graphs, nodes have a boolean attribute $inCache$, indicating their role in network evolution. The model has as parameters node degree K, minimum degree L, and maximum degree U. The graph starts with K nodes with attribute $inCache = True$. Each of these nodes has L edges incident on them from randomly chosen nodes within the group. When a new node N is introduced into the graph its $inCache$ value is $False$,

---

[5] All graph visualizations in this paper were made with the Pajek package [6].

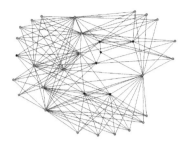 

Fig. 2. 32 Node PRU and Hypergrid Networks

and edges are added between it and L randomly selected $inCache$ nodes. If this addition causes any $inCache$ node N_C to have more than U edges, N_C has its $inCache$ value set to $False$, and a non-$inCache$ node in the system is chosen to become $inCache$ [19]. A visualization of a 32 node PRU graph is given in Figure 2 on the left with $inCache$ nodes colored black.

Hypergrid. The Hypergrid model for P2P networks, proposed by Saffre and Ghanea-Hercock [21], builds a graph topology that enforces low graph diameter and bounded node degree. The graph grows as a simple k-ary tree with nodes on the leaf level of the tree having their $k-1$ free edges randomly connected to other nodes on the same level in the tree that have degree less than k. A visualization of a 32 node Hypergrid graph is given in Figure 2 on the right.

3 Unstructured P2P Search Algorithms

Search in a graph is defined as finding a path from a randomly chosen start node N_s to a randomly chosen destination node N_d. The *cost* of a search is the number of edges traversed in locating the destination node (i.e., the number of "messages" sent between peers in the network during the search process). There are two broad classes of search techniques for unstructured P2P graphs: *uninformed* (blind) and *informed* (heuristic) [25]. Uninformed algorithms utilize only local connectivity knowledge of the graph during search. Sometimes this is the best we can do; without the ability to maintain some local state, search can do little more than follow some systematic blind routine. If we can maintain some local state, then search can proceed in a more intelligent manner. In addition to basic connectivity, informed algorithms use some localized knowledge of the graph (such as "directional" metadata) to make heuristic decisions during search. In this section we consider two characteristic uninformed search algorithms, random Breadth-First-Search (BFS) [5,9,12] and k-random walk [1,15], and a generic informed search algorithm, GAPS [26].

3.1 Random Breadth-First-Search

Random BFS [5,9,12] is an uninformed search algorithm that has been proposed as an alternative to basic uninformed BFS ("flooding"). Basic BFS is a common technique for searching graphs. Search begins at N_s by checking each neighbor for N_d. If this fails, each of these neighbors check their neighbors and this continues until N_d is found. The idea behind random BFS is to improve on the flooding method to reduce message overhead during search. This is attempted by randomly eliminating a fraction p of neighbors to check at each node. Search then proceeds from N_s with n_s neighboring nodes as follows: select $\lceil (1-p)n_s \rceil$ randomly chosen nodes adjacent to N_s, and return success if N_d is among them. Otherwise each of these neighbors randomly selects a $(1-p)$-subset of its neighbors. This process continues until N_d is located. If at any time during the search a node N contacts a "dead-end" node (a leaf in the graph), the search process backtracks to N and continues. It has recently been shown that there is an optimal value for p in certain restricted power-law networks [5].

3.2 k-Random Walk

Random walk on a graph is a well known uninformed search technique [1,15]. In this approach, a reduction in message overhead is attempted by having a single message routed through the network at random. Search proceeds from N_s as follows: randomly select one neighbor N. If $N \neq N_d$, then N similarly contacts one of its neighboring nodes, avoiding re-selecting N_s (if N has only one neighbor, it is forced to pass control back to N_s). This process continues until N_d is located. This search mechanism does not generate as much message traffic as the BFS algorithms since there is only one message being routed in the system. The trade-off is that the search response time is significantly longer. k-random walk extends this process to k random walkers that operate simultaneously with the goal of reducing user-perceived response time [15].

3.3 Generic Adaptive Probabilistic Search

As mentioned above, uninformed search is the best we can do lacking some local information. There have been several proposals to add "directional" metadata to uninformed search [4,12,26,29]. We consider here a simplification of these proposals which we call Generic Adaptive Probabilistic Search (GAPS), following the adaptive probabilistic search algorithm of Tsoumakos and Roussopoulos [26]. GAPS can be viewed as a minimally informed approach to searching in an unstructured system, making full use of the routing tables $\mathcal{T}_N = [e_1 : w_1, \ldots, e_k : w_k]$ associated with each node N. The weight w_i indicates the likelihood of successful search through neighbor N_i based on previous search results. Initially, $w_i = 1, \forall i$.

Search proceeds from N_s as follows: choose a single edge e_i from the routing table with probability $\frac{w_i}{\sum_{j=1}^{k} w_j}$, and return success if $N = N_d$ is adjacent on this edge. Otherwise, this neighbor selects one of its neighbors following the same procedure. When the destination node N_d is located, all nodes along the path from N_s to N_d (with loops removed) increment the weight in their neighbor tables for their successor in the path by 1. In this way, these nodes will be chosen with higher probability in future searches.

4   Simulation Results

To compare P2P network models in combination with search algorithms, we implemented them in a uniform framework. We considered using existing agent-based simulators [4,16], but decided that the level of implementation detail necessary for a clean investigation of topology/algorithm interaction necessitated a simple common framework. For each network of size $\mathcal{M}$ that we simulated, we used the following parameter values, which were chosen to build graphs of approximately equivalent edge count across all models:

- Random Graph: probability $p = \frac{2\mathcal{M}\log\mathcal{M}}{\mathcal{M}(\mathcal{M}-1)}$
- CAN: dimension $d = 3$
- Chord: edges $k = \log\mathcal{M}$
- PRU: $inCache$ node count $K = \frac{\mathcal{M}}{4}$, lower bound $L = \log\mathcal{M}$, upper bound $U = 3L + 3$
- Hypergrid: degree $k = 2\log\mathcal{M} + c$, for constant $c < 6$.

Table 1. Statistics of Simulated Networks

Model	# Nodes	# Edges	Avg. Degree (min/max)	Avg. Distance	Diameter	Clustering Coefficient
Random	1024	10240	20.0 (7/34)	2.65	4	0.02
PRU	1024	10350	20.21 (10/34)	2.89	5	0.25
Hypergrid	1024	10239	20.0 (2/25)	3.71	5	0.124
CAN	1024	9524	18.60 (4/45)	4.85	10	0.50
Chord	1024	9728	19.0 (19/19)	3.45	5	0.16

4.1   Topological Properties

As briefly discussed in Section 1, P2P models and MAS are anticipated to grow small world networks that also possibly have power-law degree distributions

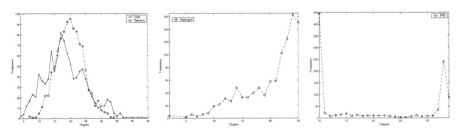

Fig. 3. Degree frequency distributions for CAN and Random model (left), HyperGrid model (center), and PRU model (right)

[3,5,10,11,18]. The results of our simulating the models under consideration for $\mathcal{M} = 1024$ are presented in Table 1. We measured these values using the Ucinet package [7]. Here, the *average distance* for a graph is the length of the shortest path between two nodes averaged over all node-pairs in the graph. The *diameter* of a graph is the length of the longest direct path in the graph between any two nodes. The *clustering coefficient* of a graph is the proportion (averaged over all nodes) of nodes adjacent to a particular node that are also adjacent to each other [28]. The node degree frequencies for the models are plotted in Figure 3.

4.2   Search Performance

We now describe our experimental setup for measuring search performance. We were interested in the actual number of edges traversed to find a node in the system. The studies discussed in Section 1.1 have primarily considered the *probability* of successful search. We were looking at the *cost* of 100% success for each

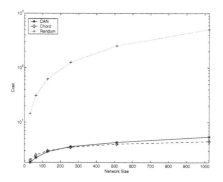

Fig. 4. Search performance comparison of structured models (CAN, Chord) using their native search algorithms against an unstructured model (Random) using BFS

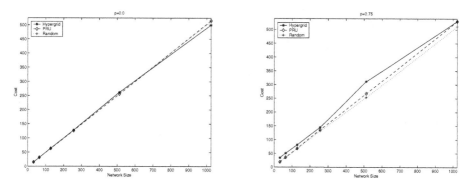

Fig. 5. Random BFS search performance across Hypergrid, PRU and Random models. Cutoff probability = 0.0 (left) and 0.75 (right).

search (i.e., Time To Live, TTL = ∞). We measured search cost, on simulations of network size 2^n for $5 \leqslant n \leqslant 10$, as the average of 5000 searches on each size (specifically: 50 simulated networks, 100 searches on each, for all 6 network sizes). For measurements of the GAPS algorithm, we "weighted" some fraction P of nodes in the system more heavily (i.e., $P\%$ of the nodes are "popular") to be the destination for some fraction W of the searches. We skewed search in this manner since the general efficacy of GAPS is dependent upon there being popular nodes in the system that are the destination nodes for a higher than average proportion of the searches. We also "primed" the network with 100 messages before measuring GAPS cost so that we could distinguish its behavior from random walk. The results of our simulations are presented in Figures 4 – 9.

5   Discussion

As mentioned above, the defining characteristics of a small-world network are low diameter and high clustering coefficient [28]. The values in Table 1 clearly indicate that all of the models (except the random model) grow small-world topologies. Chord, with a constant degree distribution, does not exhibit a power law. None of the degree distributions plotted in Figure 3 follow power-laws: CAN (left) follows a Poisson distribution (like the random graph) because it is built by assigning nodes in the graph using a uniform hash function [20]. In the case of Hypergrid graphs (center), the bulk of the nodes have maximum degree while some linearly decreasing number of nodes at the leaf level fail to establish maximum degree. PRU (right) has a highly skewed distribution: the "bump" at degree 10 represents the lower bound L on degree, while the peak at degree 33 represents nodes that have reached the upper bound U on degree. There are a nontrivial number of nodes with degree 34. These nodes were allowed to have $U + 1$ neighbors to handle an error condition in the PRU growth protocol [19]. The few intermediate nodes with degree between L and U are those currently *inCache*.

Turning to performance, Figure 4 illustrates the value of structure: the CAN and Chord native search mechanisms give $O(\log \mathcal{M})$ search performance. The cost of BFS on random graphs (typical of the unstructured models) increases linearly with network size $\mathcal{M}$, with cost roughly $\mathcal{M}/2$. Clearly, the native search mechanisms of structured networks outperform, by several orders of magnitude, flooding search on unstructured networks.

Next, we compare the three search algorithms for unstructured networks. The results for BFS with 0.0 and 0.75 cutoff values is given in Figure 5, for 1 and 16-random walk in Figure 6, and for GAPS, with 5% of the nodes popular receiving 75% of search requests, in Figure 7 (left). Clearly, all variants of random BFS have the same cost (indicating that randomness does not enhance basic BFS) and have lower cost than both GAPS and k-Random Walk. Also, GAPS has lower cost than the Random Walk search algorithm. The long term performance improvement of GAPS algorithm for the Random Graph model is presented in Figure 7 (right). Clearly this algorithm improves over time (albeit at a very

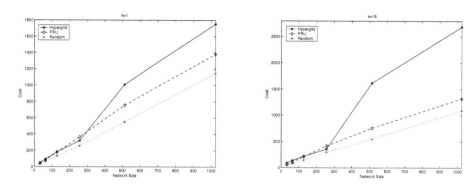

Fig. 6. k-Random Walk search performance across Hypergrid, PRU and Random models. k = 1 (left) and 16 (right).

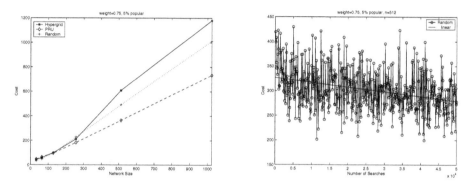

Fig. 7. GAPS (weight = 0.75, popularity = 5%) search performance (left). GAPS search performance over time, Random graph (right).

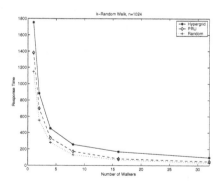

Fig. 8. k-Random Walk normalized cost (User Response Time = Cost/Number of Walkers

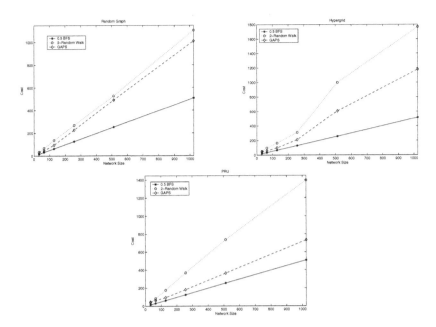

Fig. 9. Performance of search algorithms (BFS, Random Walk and GAPS) across random model (top left), Hypergrid (top right) and PRU model (bottom)

gradual rate). We also compare the user-perceived response time (that is, the normalized cost of search) of all three P2P models for k-Random Walk ($k = 1, 2, 4, 8, 16, 32$) in Figure 8. Normalized cost improvement is equivalent across all three models.

Finally, we independently consider search performance on each of the three topologies. From Figure 9, it is evident that the random graph scales well for all the search algorithms. Hypergrid has similar search cost as that of PRU and Random graph for small size networks but as the network size increases, its performance degrades. Random Walk involves the highest cost in all three graphs, making GAPS a good alternative to $k-$random walk. Overall, these experiments clearly indicate that the random graph model and BFS requires lowest cost for unstructured networks.

6   P2P Models, Search Algorithms and Learning Modules

The P2P models and search algorithms discussed and compared in this paper have recently been re-implemented in Java and integrated into the IVC Software Framework in the InfoVis Cyberinfrastructure under development in the School of Library and Information Science at Indiana University.[6] The IVC Software Framework enables non-programmer users to run diverse data mining, modeling and visualization algorithms in a menu driven way.

---

[6] http://iv.slis.indiana.edu/

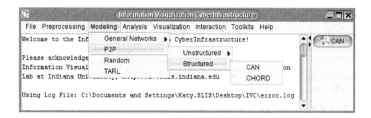

Fig. 10. Main application window of the IVC Software Framework

A snapshot of the interface to the IVC Software Framework is given in Figure 10. Continuous feedback on user requests and algorithmic results is printed in the background of the main application window. Generated networks can be analyzed using the Network Analysis Toolkit available under the 'Toolkits' menu or by running one of the diverse search algorithms under the 'Analysis' menu. Networks can be visualized using algorithms available under the 'Visualization' menu. All algorithms in the IVC Software Framework are extensively documented online. In addition, two Learning Modules are available online that aim to educate about the Error and Attack Tolerance of Networks and about the Search Performance of P2P Networks.

7   Conclusions and Future Work

In this paper we explored the topological properties and search performance of structured and unstructured P2P models using simulations of the CAN, Chord, Hypergrid, and PRU models and the random BFS, k-random walker, and GAPS search algorithms. Our goal was to provide a basis for a better understanding of the role of topology in search performance and to highlight the strengths and weaknesses of these models and algorithms.

We discovered that most of these models do indeed grow as small worlds with low diameter and high clustering coefficients. None of the models developed power-law degree distributions. We also found that basic BFS overall had lowest search cost across all unstructured models and that the random graph topology supports the lowest cost search overall using BFS. Furthermore, we determined that random cutoff does not improve the cost of BFS. We also found that increasing the number of walkers in random walk does not improve search cost; in fact, this just trades network load for user perceived response time. Finally, we found that the GAPS algorithm performs well as an alternative to k-random walk on all networks. These results indicates the need to study more closely algorithms that intelligently adapt to system dynamism and usage.

The next step in this research is to undertake a complete formal investigation of the GAPS algorithm as a paradigmatic informed search algorithm. Its generality and simplicity may give a good handle on designing efficient informed search algorithms for small-world graphs that outperform BFS. Another important step is to investigate unstructured topologies to specifically support GAPS.

Finally, an investigation of recent results which have applied percolation theory to the problem of search in power-law graphs [5,22] can profitably be pursued in our simulation framework.

Acknowledgments. We thank Beth Plale, Cathy Wyss, the reviewers, the Indiana University Database Group, the AP2PC 2004 workshop participants, and Gopal Pandurangan for their feedback and discussions on this paper. This work is supported by a National Science Foundation CAREER Grant under IIS-0238261 to the third author.

References

1. ADAMIC, LADA, RAJAN LUKOSE, AMIT PUNIYANI, AND BERNARDO HUBERMAN. Search in Power-Law Networks. *Physical Review E*, 64(4):46135-46143, 2001.
2. AKAVIPAT, RUJ, LE-SHIN WU AND FILIPPO MENCZER. Small World Peer Networks in Distributed Web Search. *Proc. ACM WWW2004*, pp. 396-397, 2004.
3. ALBERT, RÉKA AND ALBERT-LÁSZLÓ BARABÁSI. Statistical Mechanics of Complex Networks. *Reviews of Modern Physics*, 74(1):47-97, 2002.
4. BABAOĞLU, Ö., H. MELING, AND A. MONTRESOR. Anthill: A Framework for the Development of Agent-Based Peer-to-Peer Systems. *Proc. IEEE ICDCS'02*, pp. 15-22, 2002.
5. BANAEI-KASHANI, FARNOUSH AND CYRUS SHAHABI. Criticality-based Analysis and Design of Unstructured Peer-to-Peer Networks as "Complex Systems." *Proc. IEEE/ACM CCGRID'03*, pp. 351-358, 2003.
6. BATAGELJ, VLADIMIR AND ANDREJ MRVAR. Pajek: Package for Large Network Analysis. http://vlado.fmf.uni-lj.si/pub/networks/pajek/
7. BORGATTI, S.P., M.G. EVERETT, AND L.C. FREEMAN. Ucinet for Windows: Software for Social Network Analysis. Harvard: Analytic Technologies, 2002.
8. DECKER, K., K. SYCARA, AND M. WILLIAMSON. Middle-Agents for the Internet. *Proc. IJCAI97*, pp. 578-583, 1997.
9. DIMAKOPOULOS, VASSILIOS V. AND EVAGGELIA PITOURA. A Peer-to-Peer Approach to Resource Discovery in Multi-Agent Systems. *Proc. CIA 2003*, Springer LNCS 2782, pp. 62-77, 2003.
10. FALOUTSOS, M., P. FALOUTSOS, AND C. FALOUTSOS. On Power-Law Relationships of the Internet Topology. *Proc. ACM SIGCOMM*, pp. 251-262, 1999.
11. JOVANOVIĆ, M., F. ANNEXSTEIN, AND K. BERMAN. Modeling Peer-to-Peer Network Topologies Through "Small-World" Models and Power Laws. *IX Telecommunications Forum TELFOR 2001*.
12. KALOGERAKI, VANA, DIMITRIOS GUNOPULOS AND D. ZEINALIPOUR-YAZTI. A Local Search Mechanism for Peer-to-Peer Networks. *Proc. ACM CIKM'02*, pp. 300-307, November 2002.
13. KLEINBERG, JON. Navigation in a Small World. *Nature*, 406:845, August 2000.
14. KOUBARAKIS, MANOLIS. Multi-Agent Systems and Peer-to-Peer Computing: Methods, Systems and Challenges. *Proc. CIA 2003*, Springer LNCS 2782, pp. 46-61, 2003.
15. Lv, QIN ET AL. Search and Replication in Unstructured Peer-to-Peer Networks. *Proc. ACM ICS'02*, pp. 84-95, 2002.
16. MINAR, N., R. BURKHART, C. LANGTON, AND M. ASKENAZI. The Swarm Simulation System, A Toolkit for Building Multi-Agent Simulations. *Technical Report, Swarm Development Group*, June 1996.

17. MILOJIČIĆ, DEJAN S., ET AL. Peer-to-Peer Computing. *HP Labs Technical Report HPL-2002-57*, 2002.
18. NEWMAN, M.E.J. The Structure and Function of Complex Networks. *SIAM Review*, 45(2):167-256, 2003.
19. PANDURANGAN, G., PRABHAKAR RAGHAVAN, AND ELI UPFAL. Building Low-Diameter Peer-to-Peer Networks. *IEEE J. Select. Areas Commun.*, 21(6):995-1002, August 2003.
20. RATNASAMY, SYLVIA ET AL. A Scalable Content-Addressable Network. *Proc. ACM SIGCOMM*, pp. 161-172, August 2001.
21. SAFFRE, FABRICE AND ROBERT GHANEA-HERCOCK. Beyond Anarchy: Self Organized Topology for Peer-to-Peer Networks. *Complexity*, 9(2):49-53, 2003.
22. SARSHAR, NIMA, P. OSCAR BOYKIN, AND VWANI ROYCHOWDHURY. Percolation Search in Power Law Networks: Making Unstructured Peer-to-Peer Networks Scalable. *Proc. IEEE P2P2004*, pp. 2-9, 2004.
23. SHEHORY, O. A Scalable Agent Location Mechanism. *Proc. ATAL'99 Intelligent Agents VI*, pp. 162-172, 1999.
24. STOICA, ION ET AL. Chord: A Scalable Peer-to-Peer Lookup Protocol for Internet Applications. *IEEE/ACM Trans. on Networking*, 11(1): 17-32, February 2003.
25. TSOUMAKOS, DIMITRIOS AND NICK ROUSSOPOULOS. A Comparison of Peer-to-Peer Search Methods. *Proc. ACM WebDB 2003*, pp. 61-66, 2003.
26. TSOUMAKOS, DIMITRIOS AND NICK ROUSSOPOULOS. Adaptive Probabilistic Search for Peer-to-Peer Networks. *Proc. IEEE P2P2003*, pp. 102-109, 2003.
27. WALSH, TOBY. Search in a Small World. *Proc. IJCAI99*, pp. 1172-1177, July-August 1999.
28. WATTS, DUNCAN AND STEVEN STROGATZ. Collective Dynamics of 'Small-World' Networks. *Nature*, 393:440-442, June 1998.
29. YANG, BEVERLY AND HECTOR GARCIA-MOLINA. Improving Search in Peer-to-Peer Networks. *Proc. IEEE ICDCS'02*, pp. 5-14, 2002.

Distributed Hash Queues:
Architecture and Design

Chad Yoshikawa[1], Brent Chun[2], and Amin Vahdat[3]

[1] University of Cincinnati, Cincinnati OH 45221, USA
yoshikco@ececs.uc.edu
[2] Intel Research Berkeley, Berkeley CA 94704
[3] University of California, San Diego, La Jolla, CA 92093-0114

Abstract. We introduce a new distributed data structure, the Distributed-Hash Queue, which enables communication between Network-Address Translated (NATed) peers in a P2P network. DHQs are an extension of distributed hash tables (DHTs) which allow for *push* and *pop* operators vs. the traditional DHT *put* and *get* operators. We describe the architecture in detail and show how it can be used to build a delay-tolerant network for use in P2P applications such as delayed-messaging. We have developed an initial prototype implementation of the DHQ which runs on PlanetLab using the Pastry key-based routing protocol.

1   Introduction

Delay-Tolerant Networks (DTN) [1] are network overlays that enable communication even in the face of arbitrary delays or disconnections. This is accomplished by using a store-and-forward mechanism which holds packets at interior nodes until forwarding to the next hop in the route is possible. Unlike IP, there is no assumption of an instantaneous source-to-destination routing path nor are there limitations placed on latency or packet loss. In essence, arbitrary delays along the routing path are tolerated by incorporating storage and retransmission in the network itself.

DTNs are useful for enabling messaging over so-called *challenged networks* [2] which have inherent network deficiencies that prohibit communication using standard IP. Examples of challenged networks include satellite-based communication, sensor networks, and ad-hoc mobile networks.

Unfortunately, challenged networks need not be so exotic. Current trends indicate that the Internet itself is becoming a challenged network. The threat of computer virus infection has increased the proliferation and aggressiveness of Internet firewalls. In addition, the dwindling supply of public IP addresses has led to the popularity of NAT gateways which effectively hides machines behind private IP addresses [3]. In both cases, bidirectional communication has been severely constrained (by limiting port numbers) or eliminated altogether (in the case of NAT-to-NAT communication). This restriction severely limits the ability of P2P applications to make use of these NATed nodes. What we are left with is

G. Moro, S. Bergamaschi, and K. Aberer (Eds.): AP2PC 2004, LNAI 3601, pp. 28–39, 2005.
© Springer-Verlag Berlin Heidelberg 2005

a challenged network where a growing population of *private* machines can only communicate (unidirectionally) with *public* machines.

In this paper we present a solution to this problem - the distributed hash queue (DHQ). The DHQ provides durable network storage that can be used to facilitate communication between disconnected peers. A sending host places network packets into the DHQ and a receiving host subsequently pulls packets from the DHQ. All queues are named using 160-bit keys and a queue lookup (naming) service has been built on top of the Pastry key-based routing protocol. The DHQ prototype runs on top of the PlanetLab network testbed and the initial implementation consists of approximately 2500 lines of Java code.

2   Simplified DTN Architecture

As defined by [2] and [4], a general delay-tolerant network provides several different classes of service and delivery options. These include Bulk, Normal, and Expedited service and Return Receipt and Secure delivery options, among others. In addition, a DTN provides multi-hop routing across several regions using name tuples.

In this paper, we provide an implementation of a simplified DTN architecture than can be extended to the general case. Our architecture consists of a single Reliable class of messaging service and we use 160-bit hash keys for names. Delivery options are not provided by default, however, they can be added at the application level if desired. Routing is limited to single-hop paths, from a NATed network node to another NATed network node. Multi-hop paths can be built by inserting application-specific route headers into message contents but that is beyond the scope of this paper.

To summarize, then, the DTN that we describe in this paper has the following basic properties:

2-Region Connectivity. Messages can be routed between two disconnected network regions, i.e. two NATed nodes.

160-Bit Names. Message queues are named by 160-bit keys.

Reliable Delivery Option. All message are reliably delivered in the face of up to K network faults. The constant K is a configurable parameter but is set to 3 by default.

3   Background

The DHQ system makes extensive use of the Pastry key-based routing (KBR) protocol. Pastry is used to implement the DHQ name service and to help in replicating queue state. While Pastry is used for the implementation, any KBR protocol would be sufficient. In this section, we give a brief background of the Pastry system. For a complete description, see Rowstron, et. al [5].

In the most basic sense, Pastry maps 160-bit keys to IP addresses. Thus, given any 160-bit key, Pastry will return the closest IP address to that key. This provides the basis for the DHQ name service, since we need to map queue names (160-bit keys) to the host that owns the queue state.

In the Pastry system, the 160-bit key space is configured in a ring (from 0 to $2^{160} - 1$) and the nodes are distributed along the ring. All nodes are assigned a node ID which consists of a 160-bit key and a IP address. Using a consistent hashing algorithm (e.g. SHA1), the IP address is deterministically hashed to a key. In addition to being deterministic, the hashing algorithm also generally provides a uniform distribution of keys. So the nodes are roughly distributed in the 160-bit key space in a uniform manner.

An important feature of Pastry (and of other KBRs) is that the average route path from any node to the owner of an ID is $log(N)$ in the number of nodes in the system. In Pastry, in fact, the average route path is $log_b(N)$ where the base is 16. So the system can potentially scale to a large number of peers.

4   Distributed Hash Queues

The Distributed Hash Queue (DHQ) system provides a queuing service to both public and private peers on the Internet. At the highest level of abstraction, senders push messages to named queues and receivers pop messages from named queues. A request-reply messaging service can be built on top of the queuing service by using the tag field in the queue element structure to match requests with replies.

Senders and receivers are assumed to be applications running on NATed network nodes, e.g. a pair of instant-messaging applications. The DHQ service consists of N nodes running on the PlanetLab which are publicly addressable (i.e. have public IP addresses) and participate in a single Pastry ring (group of cooperating nodes). See Figure 1.

The DHQ system consists of three services: a reliable naming service, a gateway service (for accepting requests from NATed nodes), and the core reliable queuing service. See Figure 2 which shows the layered structure of the DHQ system.

4.1   Reliable Naming Service

All queue methods operate on named queues and must use the naming service in order to locate the queue owners. The naming service provides a mapping from queue names (160-bit keys) to a set of K locations which replicate the queue state for redundancy. In addition, in order to prevent the naming service itself from becoming a single point of failure in the system, names are replicated across K nodes for fault-tolerance (In practice, K is chosen to be 3). The name-to-queue-owners binding is replicated by making use of the Pastry *replica-set* feature which finds the K closest nodes to a particular ID. A queue name is first converted to a Pastry key *key*, and then the Pastry system is used to locate the K node handles which may contain the name binding.

For example, consider a lookup of the queue named "request". First, the name "request" is converted into a Pastry key $request_{key}$ which begins with the hex

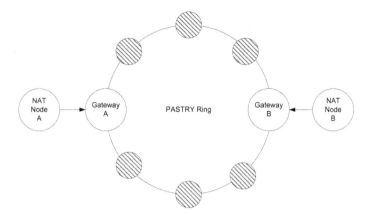

Fig. 1. This figure shows the logical structure of the DHQ service. Two communicating NATed nodes, A and B, connect to the DHQ service via the closest respective gateway node. Once connected, the NATed nodes can issue queue commands, e.g. push and pop.

digits $0x338A$.... A request message for a list of name replicas (*LookupReplicas*) is then sent to the Pastry node with ID closest to the key $0x338A$.... This closest node responds with a list of K replica node IDs. A name lookup is then attempted in parallel to each of these replicas, and the first valid response is returned to the caller. (A similar mechanism is used by the PAST storage system [6].) See Figure 3 which shows the operations involved.

4.2   Gateway Service

In the DHQ system, the NATed peer nodes do not participate in the Pastry ring, i.e. they do not own a part of the Pastry ID space. This is by design since NATed nodes are assumed to be highly dynamic and would introduce a high churn rate [7] into the system which would decrease stability. Instead, NATed nodes communicate to the Pastry ring nodes using a Gateway Protocol over standard TCP/IP. Commands are sent as human-readable single-line ASCII strings in order to ease parsing and debugging. In addition, this simple protocol makes the process of creating DHQ clients much simpler. The only requirement for a DHQ client is that it must support TCP/IP and be capable of sending ASCII strings. In fact, during the debugging process, a telnet client was used to connect to the ring and issue push and pop commands. The list of gateway commands is described below.

Alive *queue_name* lists the nodes which contain a live copy of *queue_name*. This list decreases monotonically as nodes fail until the queue is fixed using the Fix command (see below).
BlockingPop *queue_name* blocks until the queue has at least one element then returns that element.

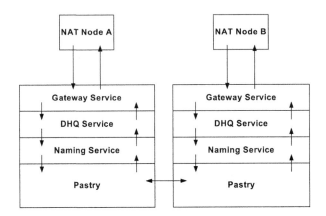

Fig. 2. This figure shows the layered structure of the DHQ system. The arrows indicate communication between layers and between entities. The DHQ and Naming services are implemented as Pastry applications and communicate strictly through Pastry. The NAT nodes connect to the DHQ service via the Gateway service which listens for TCP/IP connections.

Create *queue_name* (*IPaddress*1, *IPaddress*2, *etc.*) creates a queue named *queue_name* on the machines represented in the IP address list. In the case that no list is given, the current gateway node and its neighbors are used to replicate the queue.

Delete *queue_name* deletes a queue from the system. This removes the name *queue_name* from the naming service so that the queues are effectively deleted.

Fix *queue_name* ensures that the queue name *queue_name* is K-replicated and the queue state is K-replicated. For each queue, a Fix command is issued periodically by the system (every 2 minutes) in order to maintain the replication factor of each queue. x

Range This returns the ID space that the gateway node is responsible for. This is used for debugging purposes and to map out the distribution of the ID space to each node.

NameAlive *queue_name* returns the set of nodes that are replicating the name binding for *queue_name*. This set is not usually the same returned by the Alive command.

Peek *queue_name* returns the first element from the queue *queue_name* without removing it.

Pop *queue_name* removes and returns the first element from the queue.

Push *queue_name* **"value"** pushes an element onto the queue *queue_name* consisting of the string "value".

QueueInfo *queue_name* used for debugging. Return a string representation of the queue size and contents.

Where *queue_name* return the list of queue replicas. This is a superset of the nodes returned by the Alive command.

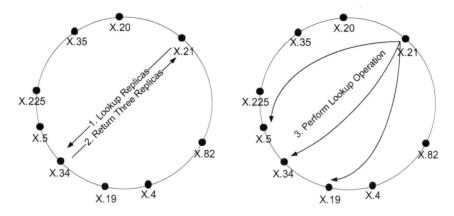

Fig. 3. This figure shows the steps taken during a lookup operation on the Reliable Naming Service. First, a set of replicas is fetched from the node closest to the name key. Then, a lookup request is multicast to these nodes and the first valid response is used.

NATed nodes attach to Gateway nodes by using a bootstrap process that is as follows. First, a NATed node contacts a seed node that it obtained via some out-of-band process. Then, the NAT node executes the nearby-node algorithm from [8] in order to find the closest Gateway node. In our experience, the nearby-node algorithm tended to be biased toward returning the seed node and an improved algorithm based on Vivaldi [9] network coordinates is currently underway.

Once the closest node is found, the NATed node opens up a socket connection to the gateway over a well-known port number. Once connected to the Gateway Service, the NAT node issues commands (one per line) and receives any responses (e.g. to pop messages) over the network stream. If a connection is lost, the Gateway can restart the bootstrap process to find a better node or try to connect directly to the Gateway again. Gateway commands are translated directly into queue operations which are then handled by the Reliable Queue Service.

4.3   Reliable Queue Service

Queues are implemented as priority queues where the message timestamps denote priority. This provides a total ordering on messages given synchronized global clocks. Given weaker time synchronization, however, the priority queues still serve a purpose: they provide a consistent ordering of packets in replicated queues. Therefore, if messages are replicated across a set of K queues, the priority feature ensures that messages will be seen by queue readers in the same order regardless of which queue is accessed.

The set of queue operations that are supported includes the set of Gateway commands plus some additional commands. Only the additional commands are listed below:

CreateQueueReplica *queue_name queue_state* This message is sent, along with serialized queue state, to a node in order to manually replicate a queue.

GetQueueState *queue_name* This command is used to fetch the entire state of a queue from a remote node.

PingQueue *queue_name* Determine if a queue exists.

WatchQueue *queue_name* This message is scheduled periodically using the Pastry *schedule − message* primitive. A WatchQueue message, when received by the QueueService, will automatically *fix* a queue and maintain the invariant that the queue and its name binding has K replicas.

Push and pop operations are multicast to all of the K queue owners in order to attempt to preserve queue consistency. For a push operation, we chose not to use a synchronous two-phase commit protocol such as ABCAST [10], but rather we use a best-effort send which attempts to send the message to all live queues. While this does not guarantee consistency, with a large enough value of K it does probabilistically guarantee that the message will not get dropped.

In order to preserve the queue state over long delays, it is important that the queues be able to survive many faults. For example, in a delay-tolerant network, days or even weeks may go by before a message can successfully be delivered. Therefore, the DHQ needs to durably store messages so that they can survive multiple faults. This is handled by the initial queue replication, and a periodic process which re-replicates queue state every S seconds. In practice, we have used a value of S to be 120 seconds, although this value is tunable and should be set according to the environment in which the DHQ is operating. In this initial implementation, the faults that we are trying to survive are mainly the periodic reboots of PlanetLab nodes. Currently, we are assuming fail-stop nodes which simplifies the implementation. Future work will be to survive other kinds of failures and to improve the consistency guarantees of the system.

5   Results

In this section, we describe the overhead and scalable performance of DHQ. In order to benchmark the system, we have built an application called *DynamicWeb* on top of DHQ. Among other things, the DynamicWeb system enables private NATed computers to serve web documents via a public queue living in the DHQ. (Note that for these performance tests, we have turned off the queue replication and reliable name services, i.e. we are testing the bare-bones DHQ performance.)

During traditional web browsing, a client initiates an HTTP REQUEST which is satisfied with an HTTP RESPONSE by the server (see Figure 4). In the DynamicWeb system, however, web browsers and web servers are indirected through the DHQ, see Figure 5. HTTP requests become DHQ *push* operations, and the resulting HTTP response is *popped* by the browser. (HTTP Requests are matched with the corresponding HTTP Responses by an auto-generated unique identifier.)

The first question to answer is, "How much does this indirection cost?", in other words how much overhead does the DHQ system introduce. In the indirect

Fig. 4. This figure shows the direct communication between web browser and web server during conventional web browsing

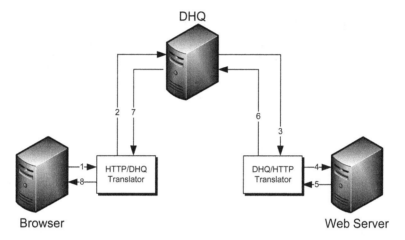

Fig. 5. This figure shows the indirect communication between web browser and web server of DynamicWeb

case, we can see from Figure 5 that there are four times as many messages as the direct case. Thus, the expected latency of the DynamicWeb request is at least four times that of the direct web request. Using a single browser, server, and DHQ node, we created a test to determine this overhead. In Figure 6, we see that the overhead is actually eight times the direct web fetch case – more than expected but still under 10ms for requesting a 16KB file. As the message size is increased past 4KB, we can start to see the linear effect of copy costs. We expect to be able to improve on this request/reply latency in the future by moving from the current store-and-forward architecture to a cut-through, byte-stream oriented architecture.

The second question to answer is, "How does the performance of the DHQ system scale?". To answer this question, we again use the DynamicWeb applica-tion as our benchmark. In this case, we have an array (15) of clients requesting one-thousand 16KB documents from an array (15) of web servers. Each client is assigned a distinct web server, all all web requests are synchronous (e.g. each client has only one outstanding request in the system at a time.) This web traffic is indirected through the DHQ system which varies in size from 1 server to 15

DHQ Overhead: Dynamic Web

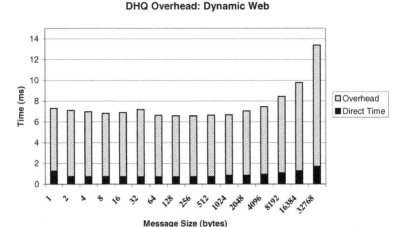

Fig. 6. This figure shows the overhead of a DynamicWeb request vs. a traditional web request, with varying request file sizes

servers. The goal of the experiment is to determine how throughput scales as more DHQ nodes are added to handle this constant load of web traffic. In order to negate any effect of load-imbalance, we have assigned DHQ nodes to the clients/servers in a round-robin manner at runtime. The results of 120 seconds of this test are shown in Figure 7.

We can see that throughput increases as more DHQ nodes are added to the system, most noticeably the two cases when 8 and 15 DHQ nodes are present. For the latter case, the throughput of the system reaches a peak of 131 requests per second. (The downward slope on this line is caused by the fact that clients are finishing their work of 1000 requests and there is not enough load on the system to sustain the peak throughput.)

We also notice that the system is not perfectly scalable. This is the result of two competing forces present: processing power and network latency. As more DHQ nodes are added, the aggregate processing power of the system increases which enables more web requests to be handled. However, as the system grows larger, network latency also increases. This increase in network latency is due to the pseudo-random placement of DHQ queues in the Pastry ring. When there is only one DHQ node, obviously all queue state is on that single node. Queue access time, therefore, is of the order of memory access time in this case. However, as the DHQ system grows larger, the probability that queue state is colocated with queue access decreases. Queue access time is now on the order of network latency, which is significantly larger than memory access latency. This effect causes the path lengths in Figure 5 to increase which causes the system throughput to decrease (since the requests are synchronous). In the future, we plan on investigating methods to decrease queue access latency, including smarter placement of queue state (e.g. locating queue state with queue access).

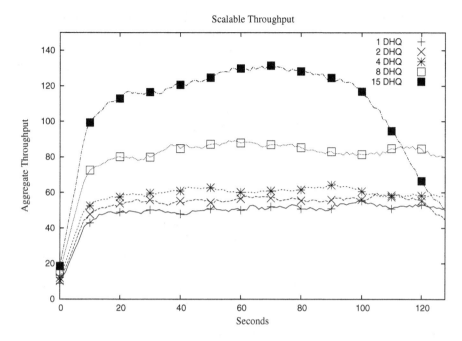

Fig. 7. This figure shows the aggregate throughput (requests/second) given varying sizes of the DHQ system

6   Related Work

In this paper, we have described a mechanism for allowing communication to a NATed network node with a private IP address. Some related work in this area has attempted to tackle this very problem including AVES [11] and i3 [12]. In AVES, the NAT gateway (and DNS server for performance reasons) is modified in order to support incoming connections to private IP hosts. A public network waypoint address serves as the virtualization of the private IP address, and relays IP packets from a public IP address to the private IP address through the modified AVES NAT gateway. The main constraint on the AVES solution is that it requires gateway software modifications which may not be administratively possible by all NATed clients. In addition, while AVES does provide general bi-directional communication from host-to-host, it still makes the assumptions of low RTT and packet loss and therefore is not a candidate for building a complete delay-tolerant network (DTN).

The Internet Indirection Infrastructure (i3) is another possible choice as a substrate for building a DTN. In i3, packets are sent not to an IP address but rather to a rendezvous node identified by an m-bit key, called k. An overlay network (Chord is used in the i3 implementation) then routes data packets to the node associated by $successor(k)$ in the Chord system. Any interested parties can register triggers with the rendezvous node (again, using the key k to identify the rendezvous node). The triggers then forward packets to the interested nodes. What i3 provides through this indirect communication is the ability for recipients to be mobile. For example, if a host moves from address 128.A.B.C to 128.X.Y.Z, then it simply must refresh its trigger to point to its new IP address. A recipient's mobility, however, is still limited to the public Internet since triggers forward packets using IP. In addition, i3 does not provide network storage for packets as is required by a DTN - packets are simply forwarded by a trigger as soon as they arrive. If the destination host is currently unavailable, then the packet is lost and must be retransmitted by the source node. In a DTN, it is the network that provides network storage and/or retransmission before failing.

The POST system [13] provides secure and reliable messaging between disconnected hosts. Like DHQ, POST is built on top of key-based routing protocol and provides message storage in the network. The main differences between the two systems is the fact that the POST system design assumes bidirectional communication between hosts (it is a P2P messaging system) and is focused on secure messaging. While end-to-end security can be added on top of DHQ at the application layer, it is not a focus of this paper.

The IP Next Layer (IPNL) system [14] provides connectivity to NATed hosts by extending IP addresses to be a triple of a public IP address, realm ID, and private IP address. Other network communication remains the same, so that the IPNL does not handle the long storage delays that are inherent to DTNs. Also, while IPNL provides a general purpose NAT-to-NAT communication mechanism, it does so by modifying the IP layer and therefore requires router modifications.

7   Conclusion

In this paper we have described the architecture and design decisions involved in building a distributed-hash queue (DHQ) service. The primary reason for building such a service is to provide rendezvous communication for private NATed peers in a P2P system. The requirements of any such service is to provide a set of public waypoints and data durability. In our system, we use the PlanetLab testbed to provide public waypoints. To survive faults and provide long-lived data durability we have chosen to replicate both the queue names and the queue state. The API of our distributed hash queue service has been described in detail and performance measurements illustrating the scalability of the system have been given. Our planned future work includes lowering the overhead of the system and analyzing its resilience to faults.

References

1. Fall, K.: Delay-tolerant networks. In: Proceedings of ACM SIGCOMM 2003, Karlsruhe, Germany (2003)
2. Fall, K.: A delay-tolerant network architecture for challenged internets. Technical Report IRB-TR-03-003, Intel Research (2003)
3. hua Chu, Y., Ganjam, A., Ng, T.E., Rao, S.G., Sripanidkulchai, K., Zhan, J., Zhang, H.: Early experience with an internet broadcast system based on overlay multicast. Technical Report CMU-CS-03-214, CMU (2003)

 4. Cerf, V.G., Burleigh, S.C., Durst, R.C., Fall, D.K.: Delay-Tolerant Network Architecture. http://www.ietf.org/internet-drafts/draft-irtf-dtnrg-arch-02.txt (2004)
 5. Rowstron, A., Druschel, P.: Pastry: Scalable, distributed object location and routing for large-scale peer-to-peer systems. In: IFIP/ACM International Conference on Distributed Systems Platforms (Middleware). (2001) 329–350
 6. Rowstron, A., Druschel, P.: Storage management and caching in past, a large-scale, persistent peer-to-peer storage utility. In: Proceedings of the eighteenth ACM symposium on Operating systems principles, ACM Press (2001) 188–201
 7. Rhea, S., Geels, D., Roscoe, T., Kubiatowicz, J.: Handling churn in a dht. Technical Report CSD-03-1299, UCB (2003)
 8. Castro, M., Druschel, P., Hu, Y.C., Rowstron, A.: Exploiting network proximity in distributed hash tables. In Babaoglu, O., Birman, K., Marzullo, K., eds.: International Workshop on Future Directions in Distributed Computing (FuDiCo). (2002) 52–55
 9. Cox, R., Dabek, F., Kaashoek, F., Li, J., Morris, R.: Practical, distributed network coordinates. In: Proceedings of the Second Workshop on Hot Topics in Networks (HotNets-II), Cambridge, Massachusetts, ACM SIGCOMM (2003)
10. Glade, B., Birman, K., Cooper, R., van Renesse, R.: Lightweight process groups in the isis system (1993)
11. Ng, T.S.E., Stoica, I., Zhang, H.: A waypoint service approach to connect heterogeneous internet address spaces. In: Proceedings of the General Track: 2002 USENIX Annual Technical Conference, USENIX Association (2001) 319–332
12. Stoica, I., Adkins, D., Zhuang, S., Shenker, S., Surana, S.: Internet indirection infrastructure. In: Proceedings of ACM SIGCOMM Conference (SIGCOMM '02). (2002)
13. Mislove, A., Post, A., Reis, C., Willmann, P., Druschel, P., Wallach, D.S., Bonnaire, X., Sens, P., Busca, J.M., , Arantes-Bezerra, L.: Post: A secure, resilient, cooperative messaging system. http://www.cs.rice.edu/CS/Systems/PAST/POST-IPTPS.pdf (2004)
14. Ramakrishna, P.F.: Ipnl: A nat-extended internet architecture. In: Proceedings of the 2001 conference on Applications, technologies, architectures, and protocols for computer communications, ACM Press (2001) 69–80

DiST: A Scalable, Efficient P2P Lookup Protocol

Savitha Krishnamoorthy, Karthikeyan Vaidyanathan, and Mario Lauria

Department of Computer Science and Engineering,
The Ohio State University
{savitha, vaidyana, lauria}@cis.ohio-state.edu

Abstract. A well-known problem found in peer-to-peer systems is how to efficiently and scalably locate the peer that stores a particular data item. In a typical formulation of the problem solution, each data item is mapped to a key; every peer stores data items corresponding to a contiguous range of keys, and locating an item requires identifying the host that holds that item's key. Here we describe Distributed Search Tree (DiST), a distributed lookup protocol based on a straightforward extension of the search tree concept. In DiST peers are assigned to groups, each group is responsible for a range of keys, and groups are located at the nodes of logical search tree. While our approach has comparable complexity to the best algorithms proposed so far (complexity is $O(logN)$), we show that its flexible design puts it at an advantage when it comes to the application of common performance enhancing techniques such as caching and replication. As an example of such advantage we describe the improvement in data lookup time and resilience obtained with key caching and table lookup replication.

Keywords: *Peer to Peer computing, DHT.*

1 Introduction

P2P (Peer-to-peer) systems tend to be highly decentralized. They typically consist of many nodes, which are symmetric in function, but unreliable and heterogeneous. The increasing popularity of peer-to-peer file sharing systems has paved a way to many interesting research problems.

The problem of finding data is at the heart of any decentralized P2P system [2][5] [11]. It is not addressed well by most popular systems currently in use, and it provides a good example of how the challenges of designing P2P systems can be addressed. The recent algorithms developed by several research groups for the lookup problem, like a central database or the DNS approach have inherent reliability, resilience and load balancing problems.

One of the challenges in a peer-to-peer network is to find the data item in a large P2P network, in a scalable manner. In this paper we present an abstract architecture, DiST with a scalable and flexible protocol for lookup in a dynamic peer-to-peer system. DiST is a distributed lookup protocol based on a straightforward extension of the search tree concept. It uses a hierarchical model with a group of peers acting as a single node of a tree. The parent of a peer belongs in the parent group. In DiST peers are assigned to groups, each group is responsible for a range of keys, and groups are located at the nodes of logical search tree. The protocol requires data items to be mapped to keys

G. Moro, S. Bergamaschi, and K. Aberer (Eds.): AP2PC 2004, LNAI 3601, pp. 40–53, 2005.

within a given range. Our lookup protocol can locate the group responsible for a data key in $O(logN/M)$ time where N is the size of the network and M *is the constant maximum members* allowed in a group. This is essentially $O(logN)$. However we show that integration of caching data along the search path, in our protocol can achieve 100% hit rates with cache sizes as small as the number of groups in the network. This is due to the fact that specific range of keys lie within a single group and caching any one key in this range implicitly acts as a cache for every key in this range. Hence, even a small cache is shown to achieve significant performance improvement.

The protocol takes care of distributing keys among the nodes (groups) of the tree in a balanced manner. This ensures that no group in the hierarchy is overloaded. Although data is uniquely present with a single group, it may be replicated among peers within the group for load balancing.

Our architecture requires every member to hold information about M other members. With small group sizes (M), the storage cost reduces significantly at the cost of increasing the load on each member of the system. DiST also has an efficient duplication mechanism by which frequently accessed keys are duplicated among members of the group. A data request may be forwarded to any of the members among whom the data is duplicated. This is more efficient since each group is only responsible for a subset of the entire range of keys.

The flexibility of DiST can be further exploited by plugging any other P2P lookup protocol within each group thus improving their scalability. For instance, if Chord [13] is used as the lookup protocol within the group, with $M = logN_{max}$ where N_{max} is the maximum peers one can expect in a network, then our protocol can perform the lookup within $O(log(N/logN_{max})) + O(log(logN_{max}))$.

The rest of the paper is organized as follows. Section II describes related work. Section III describes our architecture in detail. Section IV describes our simulation methodology and Section V discusses various results obtained.

2   Background and Motivation

Several protocols have been introduced before to provide look-up facilities for data in a Peer to Peer network. In Napster [7] architecture, clients connect to a central server that maintains the list of clients and their shared resources. It keeps track of online clients and maintains a central file index. The server becomes a single point of failure and introduces scalability issues. The centralization defies the vital requirement of decentralization in a P2P network.

Another popular protocol is that of *Gnutella* [6] [9] in which, a client broadcasts its request to its nearest neighbor after consulting its routing table. The request is broadcasted along the network for a certain allowed number of hops (Time To Live for the request). Data is not published or advertised. So queries are forwarded until they find a node that can serve the request. As the number of peers increase, the traffic required to messaging increases exponentially due to the flooding nature of the lookup.

Fast Track [3], came up with a tree structure; nodes with fast computing power and network bandwidth act as super nodes and take a server role. Clients join these super nodes and sends request to the super nodes. This architecture gives more importance to some nodes over the others, which affects its resilience. As the network size increases, the amount of overhead on these super nodes increases. The main disadvantage of such hierarchical architectures is that removal of nodes higher in the hierarchy leads to chaos.

Recently a number of look up algorithms have evolved, [13,8,10,4], which use the Distributed Hash Table (DHT) approach. *Chord* [13] uses a DHT abstraction that forms an ordered logical ring structure. It offers the capability to perform a lookup in $O(logN)$ hops. A single node maintains information about $O(logN)$ other nodes. However Chord and other similar protocols like Tapestry, Pastry [10] and Yappers [4] do not provide for arbitrariness in queries. The effectiveness of caching and replication in such architectures is not significant since a data key always maps to a unique network node.

These algorithms are effective in different ways. Many of the DHT based algorithms cannot see much benefit due to caching as data items or keys map to unique nodes. However the characteristics of requests indicate popularity of documents and hence caching and replication mechanisms can significantly impact the performance of the network. In this paper, we design a novel architecture that takes advantage of the efficiency of Distributed Hash Tables while providing flexibility that can make good use of caching and replication mechanisms. Our architecture also introduces subsystems or overlays within the network, which can make use of other key look up algorithms without disrupting the working or performance of the rest of the network.

3   Terms and Terminologies

Group: Every member in the network belongs to a group and a data lookup typically follows a hierarchy. Though a member belongs to a group, no member (peer) is important than any other member in the network and peers can join any group. The members joining a group are initially treated as buffers to replace any of the leaving members at higher levels of the tree. The group is stabilized only after the size of the group reaches a certain threshold value.

Key: Every data item available in the network is mapped to a key value using a popular hash function. Each group, a set of peers, in the network is responsible for a range of keys, a finite space (bounds which are determined by its parent group). Fig 1 shows an example of a key distribution among a set of groups in the network.

We refer to a group of peers that act as a node in a distributed search tree as a **group** and each peer within a single group as a **member**. In this paper, groups and nodes are used interchangeably. Groups may also be referred to as an individual peer in the network and will be indicated as a **network node**. The total number of peers in the network at any point is referred to as the **size of the network**.

4   DiST Architecture

In this section, we describe the design of a flexible architecture for a Peer-to-Peer system. We also enhance the basic protocol with several optimizations such as caching keys for efficient data lookup, data replication for data availability, using history information for efficient join and leave operations and exploiting features of other P2P

Table 1. Algorithm for Member Join

Formation of Search Trees:

Step 1: Node contacts a nearest neighbor with a join request
Step 2: The peer receiving the request, checks if the new node has any keys lying within the range of keys that its group is responsible for
Step 3: If it has and if the group size is less than the high watermark, then the new member joins the group and all the peers in the group and at least one member of the root group is informed. This step ensures that in scenarios where members join and leave the network, no time is spent on looking for the best group for the member
Step 4.1: Else, the join request is forwarded along the hierarchy to the root group, which decides the best group X, where the new peer can belong to, in such a way as to balance the tree
Step 4.2: If the best group is the root, it is directly added to the group and informed about other groups in the network
Step 4.3: Else, the root finds a group (say X) to which the new member should belong to and then forwards the request down the hierarchy to group X
Step 4.4: When the request reaches the peer in the parent group of X, it adds the new member as its child before forwarding the request to X
Step 5: All members of the group X are then informed of the new members arrival and data-key pairs and key responsibilities are exchanged.

lookup protocols owing to the flexibility of our architecture. We assume a symmetric, bi-directional, and transitive routing in the network in our architecture.

4.1 Design Overview

The underlying structure of the DiST protocol is a distributed search tree, with a group of peers acting together as a node of the tree. The root group of the tree holds information about the number of groups in the tree and the number of members in each group. This information is used to decide the potential group to which a new member joining the network should belong. The hierarchy in the tree is established by a parent-child relationship between a peer of the parent node and one or more peers in the child nodes. Currently, the architecture has been evaluated for binary search trees but can be generalized for other n-ary trees as well. For the purpose of inter-group communication, every member of the system holds the IP addresses of its parent, from the parent group and children from the child groups. The parent-child relationships are established while processing member join requests. DiST is fully distributed; no peer is more important than any other peer; Since the group to which a peer belongs to is not static, members may leave a group at any time.

DiST requires little information for join and leave operations and also for data lookup requests. In an N-sized network if the size of the group is M then DiST is seen to resolve lookups in $O(logN)$ and join requests require $O(logN)$ for communication. Each peer stores information about M other peers in the network.

4.2 Design Details

In this section we describe the join and leave operations which consequently build and shrink the tree respectively. We also describe our hashing, group management and key distribution mechanisms in more detail.

4.2.1 Tree Formation: Join Operation

Table 1 shows the steps involved in processing a single join request. A join request may be made to any peer in the network. The tree structure of the network is abstract and only used for routing purposes. New members joining the network hold certain key-data pairs and need to be informed of their group identity.

For example, say network node Y receives a join request from network node X. If X holds data-keys that Y's group is responsible for and the group can accommodate a member, X's request is processed and X joins Y's group and consequently some member of the root is informed of the addition. This step is helpful when the same node joins and leaves the network frequently. Also, with high probability it will join the same group and its data structures need not be updated as often. If the data-keys of X dont match with Y's range of keys or X does not have any data, Y simply forwards the join request to its parent recursively until the request reaches the root. Any member of the root may service this request. The root allots a group whose size is less than M to the new member and informs the allotted group of the addition. If no such group is found the root creates a new group in a balanced manner and informs the new member of the group. Any new group in the network acts only as a floating nodes until a certain minimum group size is reached. As floating members, they may be used to fill deficiencies in other groups and for load balancing purposes. This is an $O(logN)$ operation.

4.2.2 Leave Operation

Before a peer leaves the network, it can notify other members of the group, though this is not neccessary in the proposed architecture. The architecture ensures good resilience in the event of a peer going down or leaving without prior notification. This is because the data is duplicated among many members in a group. Also there are buffer nodes at any instant to make up for bursty leaves. Whenever a group size falls below a minimum threshold, the child merges with the parent and the tree shrinks. The peer leave information is updated lazily, only when a miss occurs for a data lookup request. This takes care of unstable peers that may be subsequently re-joining the system.

4.2.3 Hashing and Group Mapping

Data items being looked up are mapped to keys. A base hash function such as SHA-1 [1] assigns each data item a value that is referred to as the **key** or **data key**. These keys are assumed to be well distributed over the entire domain of values that can be hashed. This property is necessary to balance the load in the network so that servicing lookups is not skewed towards any particular group.

Fig 1 shows a representative distribution of keys among groups for keys ranging from 0 to 1000 with 90% delegation of keys to child groups. This fixed percentage delegation is used in mapping the data key being searched for to the group holding the data item.

4.2.4 Group Management

Group management in DiST is required during member join operations, wherein every member of a group is informed about a new member and key assignment. When processing join requests every member of the group is informed about the new member. This

Table 2. Algorithm for Key Lookup

Lookup for a Data-Key in a peer-to-peer network

Step 1: Peer generates a lookup request
Step 2: If data is available within the group, then the lookup returns success and ideally O (1)
cost if key is within group range then service the client
Step 3: Else, the data is forwarded to the parent, or one of the children
if key is within subtree range then
 if key ¡ group range then
  forward request to left child
 else
  forward request to right child
 end if
else
 forward request to parent
end if

Step 4: The data request is forwarded until the request reaches the responsible group. This
operation takes $O(logN)$

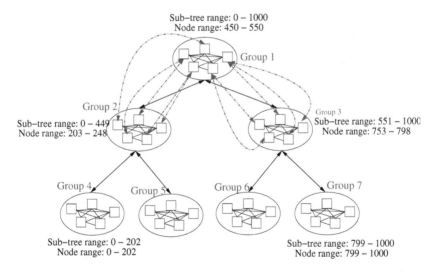

Fig. 1. Tree formation by the groups and their key responsibilities

operation requires communication with M other members, where M is the maximum
number of peers that can belong to a group and a constant of the system.

When a new group or node is formed, the parent node assigns a fraction of its data
servicing responsibility to members of its newly formed child-node. The assignment of
key responsibility to a child group requires communication with as many as M members
of the child group, where each member of the child group is updated with the range of
keys that the group is responsible for. As new groups form, the load on the parent nodes
reduces, since its data servicing responsibility is reduced. However this increases the

load in terms of request forwarding. Assignments do not occur frequently and hence this overhead incurs on a one-time basis.

4.2.5 Fault Tolerance
Key responsibilities are assigned to a group only after the group size reaches a minimum threshold value. Until a group is assigned keys, its members are treated as buffer peers. This concept of having some buffer peers provides additional fault tolerance to manage bursty leaves.

4.2.6 Data-Key Management and Lookup Mechanism
In order to search for a particular data-key in DiST, the key needs to be distributed and also be published in the network.

Key Distribution: Every data item available in the network is mapped to a key value using a popular hash function. Each group in the network is responsible for a range of keys, whose bounds are determined by its parent group. DiST distributes the keys along a balanced binary tree structure. Peers belonging to groups higher up in the tree, are assigned a lower fraction of keys or data items to service in order to balance out their forwarding responsibilities. The cost of data lookup grows logarithmically as the number of groups (depth of the tree). This scales well with large network sizes. Every member of a group is consequently notified about joins and leaves in a group. This information helps group members to monitor its size. If needed, they can take members from the child groups in order to maintain the availability of the group and also balance the load. Fig 1 shows an example of a key distribution among a set of groups in the network.

Key Publishing: When a new data item is to be stored at a peer, the group that has the key in its range is looked up. The least loaded peer in the destination group is then chosen and the data item to be stored is sent. The chosen peer adds the data item-key pair to its hash table and also updates all the peers in the group about the new key. The peer can also set an expiry time for the key, publisher's recommended expiry length for deletion of keys.

Data-Key Lookup: When a peer generates a data request, the data item is mapped to the corresponding data key value using the hashing scheme described above. Any peer of any node may receive the request. The processing of this request is described in table 2.

The member first checks if the key requested is available within its group. If it is, then look up is resolved and this corresponds to the path length of 0 in the graph shown in Fig 1. Within the group a lookup may be resolved in constant time (M being constant of the network/application using DiST). Otherwise a simple mapping algorithm (Fig 2) determines which in the hierarchy the request is to be forwarded. This may be regarded as a Distributed Hash Tree (DHT) abstraction where keys are automatically mapped to a group in the tree structure. For example in Fig 1 if group 2 gets request for a key 800, it decides to forward the request to its parent since the key does not fall in the sub-tree range. The parent peer then forwards the request to its right child and so on. However if the key happens to be within the range and no peer in the group has that key, DiST ensures that the request is aborted.

4.3   Performance Enhancement Schemes

Replication: To ensure keys remain available in the network even after the peer responsible for storing that key has failed, DiST replicates the stored keys to other peers inside a group. DiST replicates a data-key based on its key frequency, which indicates the popularity of that data item. In DiST, a particular data item gets replicated randomnly among the peers in a group. Once the data-item is replicated, it is informed to all the peers inside the group. A lookup for that key is forwarded randomly to one of the replicated peers. Also, after a series of key lookup requests for this data-item, this knowledge (replication) is disseminated among other groups with the help of caching, which is explained later in this section. In this paper, we have used replication and duplication interchangeably.

Metadata: To make searching effective, resources can be published on the network based on their metadata values (for e.g. it can be the frequency of key access). Data-items which are **popular** can be broadcasted in the network so that other peers in the network. On every key lookup request each peer can use this information and cache the results (metadata) and improve the lookup time for such popular requests.

Caching: We introduced a small cache in each member of the network to hold the key range to peer id (IP address) mapping once a lookup is resolved. A member would first consult its cache before forwarding the lookup along the tree. On a key lookup request, it first checks its cache to see whether the given key falls in the range of keys that are in the cache or close to the range of keys we are searching for. If it does, then we forward the request directly to the mapped peer. Otherwise we take the usual path along the tree. The cache holds highly distributed keys so that lookup for keys that are outside the group range can quickly reach the correct group or a nearest group, thus reducing the path length significantly.

Caching in DIST can be more effective than in any of the recent protocols [12] such as Chord. A cache in Chord would have to hold every new key that was looked up and serviced. This would require a bigger cache since a small cache that is frequently updated would be ineffective. However, DiST architecture caches only a subset of keys, which are well distributed, since and this would help to get closer to the group holding the key, even if the exact key is not cached. For example, if a key to be searched falls close to another key that was already searched, then the peer forwards that request to the corresponding member, which served that request earlier, instead of following the hierarchy. This takes advantage of the fact that a specific range of keys lies with a single group and caching any one key in this range implicitly acts as a cache for every key in this range. Caching also helps in decreasing the load on the peers that are in the top level.

For cache replacement, if a key request falls in the same range (maximum range of keys that the group holds) of a previously cached key then the peer that served this request its response replaces the old key-peer pair entry. This ensures that the cached key has the most current information. If the key request does not lie in the range of any of the cached keys then the peer serving the request replaces the least recently used key-peer pair. This ensures that the keys in cache are well distributed and also ensures that the popular keys are always in the cache. Figure 2 shows how caching strategy is applied in our architecture.

Lookups are performed in $O(logN)$ in the worst case, where N is the number of peers in the network and is shown to improve significantly with the introduction of a cache (Fig 5). If we assume that the entries of the cache are well distributed to hold a key from every group, then the key lookup moves through the hierarchy in steps of C, where C is the cache size in each member. This brings the lookup time to the order of $O(log(N/(C)))$. It is clear that when $C = N/M$, the group size, the lookup reaches the correct group in constant time, since M is a constant of the system.

As mentioned above, we need to cache only one key for every group since the responsibility of a range of keys is associated with a group and caching any one key in this range implicitly acts as a cache for every key in this range. This is analogous to memory caches where when a data item is accessed the a whole line containing it is loaded in the cache.

In this network, (N/M) is the number of groups and let C be the available cache size in each peer. The hit rate of the cache (likelihood of finding a lookup item in the cache) can be expressed as $C/(N/M)$. Now the miss rate would simple be $(1 - C/(N/M))$. Only the cache misses would incur a cost of $(logN)$ in a data lookup and constant time otherwise. The average path length of data lookup can be expressed as

Average Path Length = $C/(N/M) + (1 - C/(N/M))*logN$

Note that this expression translates into a constant lookup time when cache size is equal to the number of groups in the network. This is an interesting result, however the utilization of the cache still depends on the exact sequence of the keys requested and cache replacement policy. We have shown in the next section that the average path length is much less than $O(logN)$ with random key request patterns using the caching mechanism.

Also the average path length can be further reduced owing to the fact that a single cached key can be used to make an intelligent guess in lookup forwarding. For example, even if the requested key does not lie in the range of cached keys, the parent or child group of the cached keys could be the group responsible for the requested key. This can significantly reduce the path length.

Our simulation results (Fig 5) show that this optimal lookup time is achieved with a cache size (C) much less than the group size (N/M) for small group sizes. When groups organize themselves as a k-ary tree structure and as the network grows, we find that a key lookup would just take $O(log_k N)$ where k is the arity of the tree.

5   Simulation

In this section, we evaluate the DiST protocol by simulation. The simulation uses MPI to simulate the groups with each MPI program acting as a node of the tree. We represent these groups in a virtual group by allotting them group ids (MPI rank) and keeping track of the load on the group collectively. This is for statistics purposes only and such collective information is never maintained by the actual architecture. Our system builds a binary tree. However it can be extended to any k-ary tree. The parent and child ids of a group are established using K-ary balanced tree properties. For example, a group with id 5 in a binary tree would have groups 10 and 11 as its children and group 2 as its parent. The simulation has two main events: **Join Event and Data Lookup Event**.

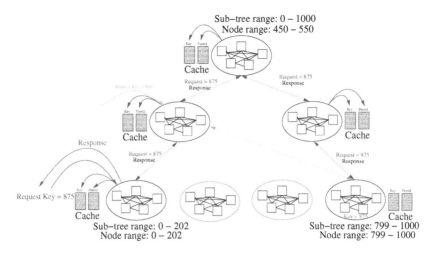

Fig. 2. Caching in DiST

Join event generates the peers using (Poisson, exponential, random) distributions with an inter arrival rate of 20ms and randomly assigns data keys to peers before joining the network. It assumes that the keys are within a fixed range (from 0 to 10000). The data lookup event generates lookup requests concurrently with join event. Members randomly add a new lookup packet to the network at a constant high rate of 10 requests / sec. At the end of the simulation we calculate the following statistics: total lookup packets serviced and forwarded; packet search time (number of hops); average network load in a group; average load/peer in a group.

A specialized scheduler process in the simulator continuously generates new member join events by randomly contacting members of the network with join requests. Members then handle the join request as described in Fig 1. Members are simulated as data structures in every MPI program. Each member has keys that it is responsible for and information about other members in its group and its child and parent ids (IP address). The simulator maintains the keys as data structure containing information about its data value, members holding the key and frequency of request for the key. A key is duplicated when the frequency of key access reaches a threshold.

The simulator uses a threshold for the maximum members (70) that a group can have. It also has a threshold for the minimum members in a group before it is assigned key responsibility by its parent. Until this minimum number is reached, these members act as the floating nodes mentioned above.

Initially every member is added to the root group, until the threshold number of members for a group is reached. Till this point, the root services all keys but the number of members requesting data is also low. When more members come in, new groups are formed and the key load is distributed. Though a group can decide how much its children's share can be, in our simulation, we have used 90% as the share that will be transferred equally to all its children.

We simulated a small cache with size ranging from 5% to 20% of the total number of groups in the network. The cache is always maintained such that the keys present are well distributed and the key information is most current.

Fig. 3. (a) Effect of number of Groups on Load in this architecture (b) Join Latency with varying group sizes for differen communication to servicing (C/S) ratios

6   Results

Arrival (join) of peers to the network is simulated by a scheduler process that contacts the nearest neighbor on behalf of the joining peer.

Fig 3a shows the decrease in load as more groups are created in the network. The load is balanced out, as the number of groups becomes 3. The root forms a balanced tree and keys are distributed to make the child groups responsible for a range of keys. When the child groups inherit responsibility of the keys from the parent, the load on the parent further reduces as shown by the figure. The load of servicing requests within a group significantly reduces as the number of peers allowed in a group and the number of groups itself increases.

Fig 3b simulates a 2^{10} sized network and evaluates the best, average and worst join latency for every 100th peer joining the network for different group sizes of 10, 30, 50 and 70. The latency was evaluated for different network communication to servicing cost ratios (C/S ration) of 0.25, 0.5, 1 and 2. The network communication was measured as the path length of forwarded join requests. The servicing cost was measured as the time taken for members within a group to update their data structure on every new member join operation.

The graphs show a clear trend towards small group sizes for faster networks and large group sizes for slower networks. This is because small group sizes lead to deeper trees and hence require more forwarding. The data structure updating is much smaller with small group sizes. However as group sizes increase, the depth of the tree decreases for a network of the same size thus decreasing the forwarding cost and increasing the inter-group communication and update of information about the new member.

For a particular C/S ratio, the latency is seen to have an optimal value of group size, below which forwarding cost dominates and servicing cost dominates above it.

The parent assigns a fraction of its keys to the child groups. Fig 4a shows the performance of different fractions of key assignment from the parent, under varying network conditions. For network communication to network node servicing time ratios varying from 0.01 to 2, the graph shows the average lookup time for key requests with varying share of keys from the parent group. Communication was measured as the number of times a request was forwarded along the hierarchy and node-servicing load as the number of requests that were served by a group member. 60% delegation of keys showed bad

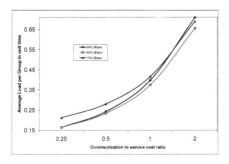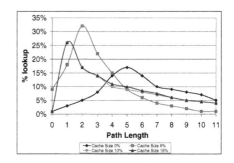

Fig. 4. (a) Effect of increase in forwarding cost on lookup time with varying parent-child share in a 2^{10} network (b) Lookup performance improvement with varying cache sizes

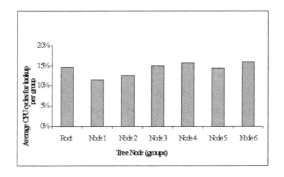

Fig. 5. Percentage of requests serviced by each group in an 8-group network

performance for all ratios. Further, we see that 80-90% delegation gives good performance in faster networks and 70-80% for slower networks. This is because with lower delegation the forwarding cost increases which dominates a slower network. The simulator currently assumes a 0.1 ratio and a static share delegation of 90% for all traces. But this can be made adaptive according to the communication to service cost ratio of the network.

Simulation results shown in Fig 5 for a 2^{10} node network, shows the average load on each group in terms of CPU cycles consumed during a lookup. For every forwarded request the cycles consumed were assumed to be one-tenth the cycles for servicing a request. This resulted in a balanced network of peers and justifies our claim to assign a large fraction of keys from the parent to every child group.

Figure 4b shows the total distribution of path lengths within which lookup requests were resolved. The performance improvement with cache sizes 0%, 5%, 10% and 15% was obtained. As in Chord [13], we define path length as the number of nodes traversed before reaching the node holding the data. When the simulator was run with 2^{10} network nodes and lookup requests totaling around 22000, with a maximum of 30 members per group, Fig 4b shows that the peak occurs as expected at $O(log(N/M))$ where N/M is the number of groups, around 70 in this run. Though an $O(1)$ lookup is expected when

C = G (70) we achieve it at C = 15 itself because of the effectiveness of caching as explained in the previous section.

For the same statistics, but with a small cache of five key-IP address pairs integrated into each member, a significant improvement is seen. This is justified because our strategy for cache management allows for widely distributed values of keys to be present in cache. A key in cache is useful for all key requests that lie within a range of keys around a cached key. The data points were obtained by running the simulator until we reached 22000 requests in the network.

7   Limitations

DiST assumes that due to its properties like buffering and group management the probability of an entire group leaving is a rarity. However such events can also be handled gracefully. The architecture can maintain a set of buffered members which can be added in the event of a complete group leaving.

8   Conclusion and Future Work

We have proposed DiST ian architecture for peer-to-peer networks with a protocol for data lookup. The architecture is based on a tree structure with a set of peers forming the node of the tree. We have described the join and lookup operations in detail and shown that these operations have a complexity of $O(logN)$ in the worst case without caching mechanisms where N is network size. For a k-ary case, DiST takes care of organizing the groups into a balanced k-ary tree. Each peer requires to store information only about the other peers in its group.

With caching, however the lookup operation is seen to give significant performance enhancements since caching any one key implicitly acts as a cache for every key in that group. We have also shown the effect of key duplication. The flexibility of the architecture allows for any protocol to be used for the lookup within a group. Hence caching, duplication and flexibility are the key features of this protocol. We believe that DiST is a practical lookup structure and will be a valuable component for peer-to-peer, large scale-distributed applications.

Our simulation assumes that group leaving is a rare event. However such events also need to be handled gracefully. Also, in DiST, active members tend to have a more updated cache than less active ones. To have uniform cache updates, caches can be integrated with timestamp information so that during node leaves and joins members can exchange caches. This will help less active members to replace stale cache entries with more recent information. In the protocol presented, there has been no attempt to address the issue of reliability to provide security, but it can be easily extended to incorporate such functionalities.

Acknowledgments

The authors would like to thank Dr. Anil Shende for his encouragement and motivation for this project. The authors also like to thank Dr. D.K. Panda for his help and support. The authors are grateful to Ms. Nagavijayalakshmi Vydyanathan, Mr. Shankar Subramanian, Mr. Radhakrishnan Sundaresan, Mr. Prashant Nikam for their guidance.

References

1. FIPS 180-1. Secure hash standard. U.S. Department of Commerce/NIST, National Technical Information Service, Springfield, VA, Apr. 1995.
2. Hari Balakrishnan and et al. Looking up data in p2p systems. In *Communications of the ACM*, February 2003/Vol.46, No.2.
3. FastTrack. Peer-to-peer technology company. In *Website http://www.fasttrack.nu/, 2001.*
4. Prasanna Ganesan, Qixiang Sun, and Hector Garcia-Molina. Yappers: A peer-to-peer lookup service over arbitrary topology. In *Stanford University.*
5. David Liben-Nowell and et al. Analysis of the evolution of peer-to-peer systems.
6. Evangelos P. Markatos. Tracing a large-scale peer-to-peer system: An hour in the life of gnutella. In *Tech Report 298.*
7. Napster. http://www.napster.com.
8. Sylvia Ratnasamy, Paul Francis, Mark Handley, Richard Karp, and Scott Shenker. A scalable content addressable network. In *Proceedings of ACM SIGCOMM 2001*, 2001.
9. Matei Ripeanu and Ian Foster. Mapping the gnutella network: Macroscopic properties of large-scale peer-to-peer systems.
10. Antony Rowstron and Peter Druschel. Pastry: Scalable, decentralized object location and routing for large-scale peer-to-peer systems. In *Microsoft Research.*
11. Mario Schlosser and Sepandar D. Kamvar. Modeling interactions in a p2p network. In *Stanford University.*
12. Tyron Stading, Petros Maniatis, and Mary Baker. Peer-to-peer caching schemes to address flash crowds. In *Stanford University.*
13. I. Stoica, R. Morris, D. Karger, M. F. Kaashoek, and H. Balakrishnan. Chord: A scalable peer-to-peer lookup protocol for internet applications. Proceedings of ACM SIGCOMM, San Diego, August 2001, pp. 160177.

A Policy for Electing Super-Nodes in Unstructured P2P Networks

Georgios Pitsilis[1], Panayiotis Periorellis[2], and Lindsay Marshall

University of Newcastle upon Tyne, U.K.
{Georgios.Pitsilis, Panayiotis.Periorellis,
Lindsay.Marshall}@ncl.ac.uk

Abstract. Unstructured P2P networks, despite having good characteristics such as the nonexistence of a single point of failure, the high levels
of anonymity in the search operations and the exemplary dependability,
have been found to be much less scalable than first expected. The flooding protocol, which is used for the discovery of peers and for the main
operation of searching, seems to be responsible for this weakness. The
adoption of some major improvements, such as the distinction between
Leaf-nodes and Ultra-Peers, has partially overcome the scalability problems, but there is still a need for further optimization. Our proposed idea,
aims to improve the effectiveness of the hierarchical scheme by applying
some new criteria in the selection of potentially promotable nodes.

1 Introduction

Gnutella is one of the most popular decentralized peer-to-peer networks. Scalability limitations have driven its development to a new two-level organization.
The new organization has its own vulnerabilities due to its dependency on the
co-operation between a relatively small number of high-level peers. Researchers
that have applied models of population dynamics to peer-to-peer systems [1],
mention the loss of confidence in users as the only weak point of this type of
networking since there are no central entities that could be forced to close down.
In our work we use graph theory and statistical analysis to determine whether
topologically important nodes have the potential for becoming Ultra-peers. We
assess topological importance using connectivity measures derived from graph
theory. We have assumed that potential Ultra-peers are peers that have to deal
with large numbers of messages. This paper tries to establish whether there
is a link between topologically important peers and the amount of traffic they
handle. In our work we have used 2 connectivity measures: Outer Degree and
Elementary Cycle, both of which we explain later. The objective is to identify
topologically important nodes and see whether they can be suggested as the
most appropriate ones, for promotion to the higher level of hierarchy, based on

---

[1] Scholar of Greek State Scholarships Foundation (IKY).
[2] Supported by the GOLD project.

G. Moro, S. Bergamaschi, and K. Aberer (Eds.): AP2PC 2004, LNAI 3601, pp. 54–61, 2005.

the traffic they generate. The rest of the paper is organized as follows: In section 2 there is a description of the problem and the solutions adopted so far, section 3 is dedicated to our solution, section 4 contains some simulation experiments and 5 the analysis of the results. Section 6 discusses the applicability of the algorithm.

2 Description of the Problem

The Gnutella protocol has a relatively simple specification [2]. The basic node discovery resource searching mechanism has its basis on a simple flooding mechanism. Messages are sent first to neighbors and upon their reception they are propagated again to their known neighbors, and so on, until the TTL (Time-To-Live) [3,4] value expires. It has been shown that nearly two thirds of the communication within the network is generated by the Ping and Pong messages. This creates a huge traffic overhead. These limitations caused by the flooding protocol itself, created the known scalability barrier of the Gnutella network, where every peer is restricted to see a certain number of other peers, which forms its horizon. Studies that are based on simulations [5] show that P2P actually scales much better than conventional theory would indicate. Thus, exponential growth of the messaging load should not be assumed. Even though we have no reason to disagree with these findings, we think that the scalability limitations always exist and radical solutions have to be deployed to shift the scalability barrier.

2.1 Some Proposed Solutions to the Congestion Problem

There have been many suggestions regarding the overhead produced by the huge number of ping messages flowing through the network. Massey [4] suggests that the maximum connectivity of a peer in the network should be restricted by the peer's bandwidth connection to the network. Other proposed solutions to the problem such as [6] are based on the idea of building and maintaining the good topological characteristics of the network. Other relevant papers such as [7] address the scaling problems via an optimized routing mechanism. The concept of Ultra-peers (or super-nodes) and the consequent separation of Gnutella in 2 layers is adopted by protocol version v0.6 [8]. Issues of reliability and efficiency were researched in [18].

2.2 Potential Weaknesses of the Existing Ultra-Peer Election Algorithm

As we stated in the previous paragraph, the employment of Ultra-peers in the Gnutella network helps the network to scale and extend its life beyond the limits that were first thought possible [9]. Allocation may be done on a voluntary basis and needs no centralized control to work but we believe there is further room for improvement. Ultra-peers and Ultra-peer candidates are usually unable to know the connectivity characteristics of their position in the network. Our aim

with the experiments and case studies presented in this paper is to assist in
building two-level networks of better connectivity characteristics by providing
some rationale behind Ultra-peer election.

3 Motivation

The motivation is to identify better election policies for Ultra-peers. We believe
that peers that deal with a large number of messages should be considered as
potential Ultra-peers. The question that we raise is 'how can the busiest peers
be identified in a dynamic network such as Gnutella ?'. The solution we propose
is to graph the peers together based on the messages they receive and use the
graph to carry out certain statistical analyses. As we show, our analysis can
identify those busy nodes without the need for physically count any messages.
We use two connectivity measures; namely the outer degree and the elementary cycle value [10] to determine important nodes based on those connectivity
measures. Then we compare the highly connected peers against their message
queues in order to determine if there is a link between topologically important
peers and traffic. We finally show from the case studies that peers selected as
highly connected in terms of elementary cycle value are also likely to attract a
lot of traffic. We acknowledge that a high connectivity factor doesn't necessarily
mean a privileged position within the graph. This is because the connectivity
factor that is given to a node when it is created is more or less a static property
that is not affected by the general position of the node in the graph.

3.1 The Connectivity Measures - Terminology

Before we examine the results of the case studies let us define the two basic
connectivity measures that we use.
Outer degree: This is a connectivity measure. The Outer Degree of node i indicates how many nodes are connected to i. Let a_{ij} be the adjacency matrix where
$a_{ij} = 1$ if i is connected to j and $a_{ij} = 0$ if i is not connected to j. Let n be the
total number of nodes. The Outer Degree is estimated by

$$OuterDegree_i = \sum_{j=i}^{n} A_{ij} \qquad (1)$$

Elementary cycle [14] is a measure of the participation of all nodes of a network in certain structures such as triangle or cycle formation. One of the structures we considered is the elementary (minimal) cycle structure; the triangle. A
cycle is a sequence of vertices of the form $C_i = (x_{i_0}, x_{i_1}), (x_{i_1}, x_{i_2}), \ldots, (x_{i_{r-1}}, x_{i_r})$
where $x_{i_0} = x_{i_r}$ (i.e. the initial vertex of the path is also the terminal vertex of
the path). The path is elementary (simple), if it does not transverse any node
more than once.

3.2 Our Approach

Many papers [10,11,12] have acknowledged the fact that Gnutella does not scale
well. We find this partially true or at best dependent on the connectivity measure

selected. The majority of the work mentioned earlier is based on the Outer degree. Other connectivity measures however have revealed a scale-free topology. Our approach is based on the hypothesis that certain connectivity values (e.g. elementary cycle) of any sub-graph of Gnutella reveal a scale-free topology (i.e. uneven distribution of connectedness) [13,16,17]. Later in the paper we show through our case studies that the top peers in terms of elementary cycle value are in fact the busiest in terms of traffic they handle. Our comparative analysis against current selection policy (which is based on Outer degree) has revealed that our method has a higher chance of spotting a potential Ultra-peer (based on how much traffic it handles) as opposed to how connected in terms of Outer Degree it is.

3.3   Simulation

For our case studies we assumed several 100 node networks. To show the correctness of our hypothesis we ran a total of 10 simulations of Gnutella communities which were randomly developed. All simulation scenarios were set up in such a way that the connectivity factors allocated to the nodes would follow power law distribution with exponent -1.4 [17]. Each node was given a unique ID number and we let the simulation to run for about 5000 virtual clock ticks. A clock tick should be seen as the interval of virtual time that is needed for an elementary processing task to be carried out on a message. We use that notion in our experiments to simulate the variation in connection speeds that can be found in a real situation and drive to congestion. For the simulation we used a tool that we built in Java for the purpose of our experiments and can emulate the Gnutella protocol version 0.4. The configuration we used in the simulation were set to the following values as we tried to make the simulations as realistic as possible:

- Ping Frequency: 1 ping send out every 30 units of virtual time.
- Probability of using info received from pongs: 50%
- Probability of ping forwarding: 60%
- Time To Live factor in the message forwarding: 5
- Connectivity: Follows Power law distribution with exp -1.4

Next we analyze the simulated graphs by using the algorithm we presented in the previous section.

4   Case Studies

In all case studies we present in this section, we show the benefit of our algorithm by comparing it with two other selection policies. In the following tables we display the 10 most highly connected nodes in terms of Outer Degree, the best 10 in terms of *Elementary Cycle value* and a random choice of peers selection. The *NodeID* that appears in every table is the unique identifier of the peer and the *Traffic* value presented in the third column shows the traffic measured in the selected peer. In the simulated protocol as traffic we consider the number of messages (pings-pongs) generated by the examined peer itself, as a result of its own discovery needs. In total, we ran 10 case studies, but due to space limitations we present results from only 2 of them. The cumulative results, however, have been drawn from all 10 studies.

Case Study 1. The two tables correspond to the top peers that have been selected in terms of their Outer Degree and in terms of Elementary Cycle. Our method has found 8 peers from the top 10 of traffic, opposed to 6 of the Outer Degree method.

Selection based on connectivity

Node ID	Outer Degree	Traffic	On Top 10
22	6	54140	√
15	5	35182	
1	5	9925	
26	5	24877	√
36	5	22212	
58	5	9006	
62	5	18107	√
75	5	18024	√
80	5	17526	√
83	5	17397	√
Total	Over 10 Nodes		6/10 or 60%

Selection based on our formula

Node ID	Elem. Value	Traffic	On Top 10
22	9	54140	√
80	9	17526	√
76	8	24877	√
75	8	18024	√
62	7	22212	√
1	7	9925	
26	7	17397	√
77	7	17526	√
58	7	9006	
81	7	25002	√
Total	Over 10 Nodes		8/10 or 80%

Comparison chart

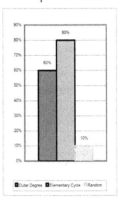

Case Study 2. Peers in high elementary cycle value handled most traffic. The comparison section in the case studies illustrates the success rate of each selection method in finding potential Ultra peers in terms of traffic. In all case studies our method appears to have about the best performance.

Selection based on connectivity

Node ID	Outer Degree	Traffic	On Top 10
16	6	25792	√
1	5	5345	
30	5	11427	
31	5	9359	
39	5	26681	√
42	5	27426	√
62	5	32038	√
67	5	4478	
76	5	29410	√
79	5	7399	
Total	Over 10 Nodes		5/10 or 50%

Selection based on our formula

Node ID	Elem. Value	Traffic	On Top 10
16	9	25792	√
63	9	24321	√
71	9	24261	√
74	9	24455	√
76	9	29410	√
62	8	32028	√
42	7	27426	√
79	7	7399	
Total	Over 10 Nodes		7/10 or 70%

Comparison chart

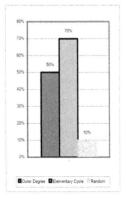

5 Results

In the next figure we show the advantage of our method in selecting the busiest peers against an Outer degree-based selection policy. "Value" denotes the number of peers that belong to the top 10 in terms of traffic. As the results show, our method always does better in selecting the high traffic nodes and on average is more successful by an average of 15.12%. In other words, picking a peer that our algorithm suggests is more likely by 15.12% to be one of the top 10 busiest ones than choosing it by its Outer Degree. Due to low dispersal in the connectivity and elementary values, in some case studies shown in the table below, we present the top 12 or top 7 of distinguishable nodes and the results have been normalized to correspond to top 10 classification. Given that in a Gnutella network the

Case study	Outer Degree selection		Elementary cycle	
	value	%	value	%
1	6	60	8	80
2	5	50	7	70
3	6	60	8	80
4	10 out of 16	62.5	12 out of 18	66
5	6	60	7 out of 11	63.6
6	7 out of 12	58.3	11 out of 18	61.1
7	7 out of 11	63.6	5 out of 7	71.4
8	8 out of 13	61.5	5 out of 8	62.5
9	6	60	5 out of 7	71.4
10	5	50	7 out of 12	58.3
Average		58.6%		68.4%

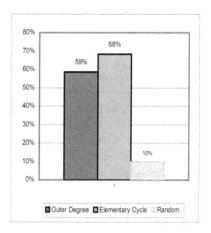

percentage of peers acting as Ultra-peers is less than 10 percent of the whole community, elementary cycle can be a more effective criterion in spotting the best amongst the whole community.

6 Applicability

As it can be seen from the description in paragraph 3.3 our algorithm requires that we have a global picture of the infrastructure to form a decision. This picture should exhibit the participation of every peer in a loop formation expressed in gathered topological data. The solution of using topology data crawlers [15] to collect such data might look easy and applicable but carries the known problems and weaknesses that centralized solutions have. Alternatively, the peers themselves could collect that information by exploiting the information communicated by their neighbors. In order to apply our technique in a typical Gnutella P2P network we need to look at the protocol level and in particular into the node discovery architecture. We need to distinguish between pong messages received by a neighbor as opposed to pong messages received by a peer on the recommendation of a neighbor. Let us call these latter type of ping messages r-pong (from recommended pong). By keeping count of the r-pong messages a peer can estimate the number of elementary cycles it participates in. Let us say for example that we want to determine the position of peer A in relation to B and C. If any pong messages meet the following criteria, then A belongs to the triangle ABC. The criteria that have to be met for a pong message to be characterized as r-pong are:

- pong has hop count = 2
- pong has been received from the connection with C (or B)
- pong carries the identity of B (or C respectively)

In networks that run Gnutella protocol, the information required for these three steps can be found attached on the message descriptors. The higher the number of different r-pong messages a peer is receiving (in comparison with the rest of a sub-network), the higher the possibility of being a potential Ultra-peer. Assuming an honesty-based type of cooperation between the peers (as they need to supply their personal data to the community), the problem is how to enforce this type of policy on a network of autonomous entities. Such a policy requires that all the involved peers would have agreed to all decisions that may have been reached within the community about sharing the *Elementary Cycle* information. Matters such as the actual selection process and consequent promotion of peers are out of the scope of this paper. Comparing our method with other based on connectivity statistics taken from the derived measured traffic, we suggest ours as more convenient since it simply requires shorter time to distinguish participation in a triangle from the r-pongs and thus decisions for promotions can be made quicker. Besides, as the simulation shows, elementary cycle has a higher chance of spotting a potential Ultra-peer.

7   Conclusion

Scale-free networks exhibit a number of properties that distinguish them from random networks. These properties along with other can be used to distinguish between important nodes and less significant ones. Gnutella on the other hand does not exhibit scale free properties that could help us define an election policy for Ultra-peers based on topologically important nodes. In this paper however we showed that other measures of connectivity can yield some topological characteristics which can be used to draw conclusions about certain peers. We took 10 Gnutella neighborhoods and analyzed them in terms of Outer Degree and elementary cycle value. We showed that there is a strong link between peers high in elementary cycle value and also in terms of traffic they handle. We propose this method to be used as an election policy for suggesting Ultra-peers.

Acknowledgments

We would like to thank Apostolos Niaouris for his contribution in clarifying our mathematical definitions. All the data and case studies as well as extensive report on this topic, are available on request from the Authors.

References

1. A. H. Chen and A. M. Schroeder: *"A Modified Depensation Model of Peer to Peer Networks: Systematic Catastrophes and other potential Weaknesses"*. AMATH 383 (June 2002).
2. The Gnutella v0.4 protocol Specification, http://www.clip2.com
3. D. Zeinalipour, T. Folias: "A quantitative Analysis of the Gnutella Network [4] Traffic". University of California - Riverside, Dept. of CS & Engineering (June 2002).
4. R. Massey, S. Bharath, A. Jain:, "Gnutella-Pro:What bandwidth barrier?", http://www.cs.ucsd.edu/classes/wi01/cse222/projects/reports/gnutellapro-5.pdf
5. R. Schollmeier,I. Schollmeier: "Why Peer-to-Peer Does Scale:An analysis of P2P Traffic Patterns". In Proceedings of *2nd International Conference on Peer-to-Peer Computing P2P'02*, IEEE.
6. G. Pandurangan, P. Raghavan and E. Upfal: "Building Low-Diameter Peer-to-Peer Networks". *IEEE Journal on Selected Areas in Communications (JSAC)*, 21(6), (August 2003) , 995-1002.
7. M.Prinkey: "An Efficient Scheme for Query Processing on Peer-to-Peer Networks". Aeolus Research Inc., Technical report (August 2002).
8. C. Rohrs, A. Singla: "Ultrapeers: Another Step Towards Gnutella Scalability", (November 2002), http://www.limewire.com/developer/Ultrapeers.html
9. Ernest Miller: "Compulsory Licensing - The Death of Gnutella and the Triumph of Google", Yale Law school web site, http://research.yale.edu/lawmeme/
10. P. Periorellis et. al. : "Dealing with Complex Networks of Process Interactions: A Security Measure". In Proceedings of *the 9th IEEE International Conference on Engineering of Complex Computer Systems*, Florence, Italy (14-16 April, 2004) .
11. J. Ritter: "Why Gnutella Can't Scale. No, Really", (February 2001), http://www.darkridge.com/ jpr5/doc/gnutella.html
12. R. Albert, A. L. Barabasi: Statistical mechanics of complex networks. Reviews of Modern Physics, (2002), 74(1): p. 47-97.
13. A. L. Barabasi, R. Albert: Emergence of scaling in random networks. *Science* vol.286 (1999) p.509-512.
14. A. Carre: Graphs and Networks, Clarendon Press, (1979), Oxford.
15. D. Zeinalipour-Yazti, M. Dikaiakos: "Design and Implementation of a Distributed Crawler and Filtering Processor". In Proceedings of *the fifth Workshop on Next Generation Information Technologies and Systems (NGITS 2002)*, vol.2382, Springer (June 2002) p.58-74.
16. A. Moura, Y. Lai, A. Motter: "Signatures of small-world and scale-free properties in large computer programs", Physical Review. E 68,017102 (2003)
17. M. Jovanovic, F. S. Annexstein and K. A. Berman: "Modeling Peer-to-Peer Network Topologies through Small World Modelsand Power Laws", In TELFOR, Belgrade, Yugoslavia, (November 2001).
18. B. Yang, H. Garcia-Molina: "Designing a Super-Peer Network". In Proceedings of *the 19th International Conference on Data Engineering (ICDE)*, IEEE Computer Society (5-8 March 2003) Bangalore, India.

ACP2P: Agent Community Based Peer-to-Peer Information Retrieval

Tsunenori Mine[1], Daisuke Matsuno[2], Akihiro Kogo[2], and Makoto Amamiya[1]

[1]Faculty, [2]Graduate School
of Information Science and Electrical Engineering,
Department of Intelligent Systems, Kyushu University,
6-1 Kasuga-koen, Kasuga, Fukuoka 816-8580, Japan
{mine, kogo, amamiya}@al.is.kyushu-u.ac.jp,
http://www-al.is.kyushu-u.ac.jp/~mine/mine-e.html

Abstract. This paper proposes an agent community based information retrieval method, which uses agent communities to manage and look up information related to users. An agent works as a delegate of its user and searches for information that the user wants by communicating with other agents. The communication between agents is carried out in a peer-to-peer computing architecture.

In order to retrieve information relevant to a user query, an agent uses two histories : a query/retrieved document history(Q/RDH) and a query/sender agent history(Q/SAH). The former is a list of pairs of a query and retrieved document information, where the queries were sent by the agent itself. The latter is a list of pairs of a query and the address of a sender agent and shows "who sent what query to the agent". This is useful for finding a new information source. Making use of the Q/SAH is expected to have a collaborative filtering effect, which gradually creates virtual agent communities, where agents with the same interests stay together. Our hypothesis is that a virtual agent community reduces communication loads involved in performing a search. As an agent receives more queries, then more links to new knowledge are acquired. From this behavior, a "give and take"(or positive feedback) effect for agents seems to emerge.

We implemented this method with Multi-Agent Kodama, and conducted experiments to test the hypothesis. The empirical results showed that the method was much more efficient than a naive method employing 'multicast' techniques only to look up a target agent.

1 Introduction

The rapid growth of the World Wide Web has made conventional search engines suffer from decreasing coverage in searching the Web. Internet users meet information floods every day, and are forced to filter out and choose the information they need.

In order to deal with these problems, a lot of studies on distributed information retrieval(e.g. [1]), information filtering(e.g. [2]), information recommendation (e.g. [3]), expert finding(e.g. [4]), or collaborative filtering (e.g. [5],[6],[7],[8])

G. Moro, S. Bergamaschi, and K. Aberer (Eds.): AP2PC 2004, LNAI 3601, pp. 62–73, 2005.

have been carried out. Most systems developed in that research are, unfortunately, based on the server-client computational model and are often distressed by the fundamental bottle neck coming from their central control system architecture. Although some systems based on the peer-to-peer (P2P for short) computing architecture (e.g. [9],[10],[11],[12]) have been developed and implemented, each node of most those systems only deals with simple and monolithic processing chores.

Considering these issues, we presents an Agent Community based P2P information retrieval method (ACP2P method for short), which uses agent communities to manage and look up information related to a user query. An agent works as a delegate of its user and searches for information that the user wants by communicating with other agents. The communication between agents is carried out based on a P2P computing architecture. In order to retrieve information relevant to a user query, an agent uses two histories : a query/retrieved document history(Q/RDH for short) and a query/sender agent history(Q/SAH for short). The former is a list of pairs of a query and retrieved document information, where the queries were sent by the agent itself and the document information includes the addresses of agents that returned the document. The latter is a list of pairs of a query and a sender agent's address and shows "who sent what query to the agent". This is useful for finding a new information source. Making use of the Q/SAH is expected to have a collaborative filtering effect, which gradually creates virtual agent communities, where agents with the same interests stay together. Our hypothesis is that a virtual agent community reduces communication loads involved in performing a search. As an agent receives more queries, then more links to new knowledge are acquired. From this behavior, a "give and take"(or positive feedback) effect for agents seems to emerge. We conducted the experiments to test the hypothesis, i.e., to evaluate how much the Q/SAH work for reducing communication loads and for causing a "give and take" effect. The experimental results showed that the method reduced communication loads much more than other methods which do not employ Q/SAH to look up a target agent, and was useful for creating a "give and take" effect.

The remainder of the paper is structured as follows. Section 2 considers the ACP2P method. Section 3 discusses the experimental results and Section 4 describes related work.

2   Agent Community Based Peer-to-Peer Information Retrieval Method

2.1   Overview of the ACP2P Method

The ACP2P method employs three types of agents: user interface(UI) agent, information retrieval(IR) agent and history management(HM) agent. A set of three agents (UI agent, IR agent, HM agent) is assigned to each user. Although a UI agent and an HM agent communicate only with the IR agent of their user, an IR agent communicates with other users' IR agents not only in the community

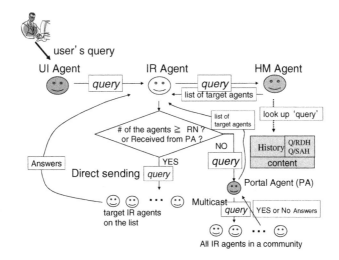

Fig. 1. Actions for Sending a Query

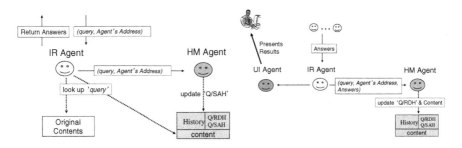

Fig. 2. Actions for Receiving a Query(left) and Answers(right)

it belongs to, but also in other communities, to search for information relevant to its user's query. A pair of Q/RDH and Q/SAH histories is managed by the HM agent.

Fig. 1 show the processes or data flows in the cases that an IR agent sends a query. Fig. 2(left and right) show in the cases that an IR agent receives a query from another IR agent or a portal agent, and an IR agent receives answers from other IR agents, respectively. When receiving a query from a UI agent, an IR agent asks an HM agent to look up target agents with its history or a portal agent to do it using a query multicasting technique (Fig. 1). When receiving a query from other IR agents, an IR agent looks up the information relevant to a query, sends an answer to the query sender IR agent, and sends a pair of a query and the address of the query sender IR agent to an HM agent so that it can update Q/SAH (Fig.2, left). The returned answer is either a pair of a 'YES' message and retrieved documents or a 'No' message representing no relevant information, provided that retrieved documents are not returned when the query comes through a portal agent. When receiving answers with a 'YES'

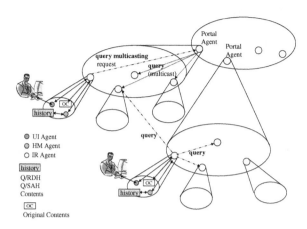

Fig. 3. Agents and their Community Structure

message from other IR agents, an IR agent sends them to a UI agent, and sends them with a pair of a query and the addresses of answer sender IR agents to an HM agent (Fig.2, right). Fig. 3 shows an example of the agent community structure which the ACP2P method is based on.

A portal agent in the figure is the agent which is a representative of a community and manages all member agents' addresses there, where a member agent of a community designates an IR agent. When a member agent wants to find any target agents which have information relevant to a query, the agent looks them up using two histories: Q/RDH and Q/SAH, and Content files. If the target agents are found, a query is sent directly to them, and their retrieved results are also returned directly to the query sender IR agent. If the requested number of such agents is not found, the agent asks the portal agent to send the query to the all member agents in the community by a multicast technique. At that time, all the answers will be returned to the portal agent. If the number of results with a 'YES' message reaches the requested number, without waiting for the rest of answers by other IR agents, the portal agent sends them back to the query sender IR agent. Even if the number of 'YES' messages did not reach the requested number after all IR agents replied, the portal agent also sends the currently held results to the query sender IR agent.

2.2   Document Content and Histories

Table 1 shows the formats of a document content file: **Content** and two histories: **Q/RDH** and **Q/SAH**. The document content file consists of a list of 4-tuples <**title, body, original, range**>, namely the title of a retrieved document, its text content, the address of the IR agent whose user owns the document, and the allowed distribution range of the document, respectively. All documents retrieved and returned by other IR agents are shared into the Content file without any redundant registration.

Table 1. The structures of Content file, Q/RDH file, and Q/SAH file

Content	title	the title of document
	body	the content of document
	original	the address of the IR agent whose user created the document
	range	the range allowed to be distributed(ALL, Community, Agent)
Q/RDH	query	a query sent by the IR agent itself
	from	the address of other IR agent which has replied to the query in the query field
Q/SAH	query	a query sent by the agent recorded in the from field
	from	the address of other IR agent who sent the query in the query field

Table 2. A part of document content

title	body	original	range
Netscape informal FAQ Japanese version	HTML text in the file	com_Netscape@	ALL

Table 3. A part of Q/SAH

query	from
telegram	root.p2p.com_telegram@
treatment	root.p2p.sic_hepatitis_type_C@
Asthma	root.p2p.sic_Asthma@
Human	root.p2p.sic_Adult_Children@
Thing	root.p2p.sic_Alzheimer's_Disease@
Ill	root.p2p.sic_Jacob_Disease@
Dream	root.p2p.sic_Dealer@
Mastocarcinoma	root.p2p.sic_Mastocarcinoma@
Hoof	root.p2p.sic_Hoof-and-Mouth-Disease@

The Q/RDH file comprises a list of pairs of <**query, from**>, each of which is a query sent by the agent itself and the address of IR agent that returned this retrieved information, respectively. The Q/SAH file is a list of pairs <**query, from**>, each of which is a query and the address of the agent which sent the query to the IR agent. Table 2 shows an example of part of a document content file. Table 3 also shows an example of part of Q/SAH file, which was originally written in Japanese.

2.3   Determining Target Agents Using Two Histories

In order to determine the target agents to send a user query, the IR agent uses the contents of retrieved document files and two histories, Q/RDH and Q/SAH. Fig. 4 depicts an example how the target agents are found, where Ⓐ to Ⓔ represent IR agents. For simplicity, we assume here that the IR agent does the

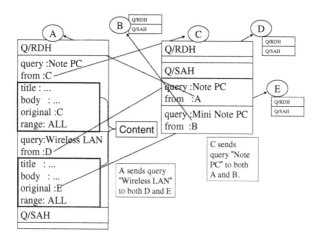

Fig. 4. Example to find target IR agents using two histories. A, B, C, D and E in circles represent IR agents' name, respectively.

job of an HM agent. Furthermore, to show the correspondence between a query and a retrieved document, we show the content file in Q/RDH.

Ⓐ has two query entries in its Q/RDH. Both queries were sent by Ⓐ itself. This figure shows that Ⓐ sent query 'Note PC' to Ⓒ and got the retrieved results from Ⓒ. Ⓒ recorded the query and Ⓐ's address into its Q/SAH. Since Ⓐ received the results from Ⓒ, Ⓒ's address was recorded in the 'from' field of the same record as the query in Ⓐ's Q/RDH. In addition, since the content included in the results is the original of Ⓒ's user, Ⓒ's address is seen in the 'original' field of the content. In the same way, Ⓐ also sent query 'Wireless LAN' to Ⓓ, Ⓓ returned retrieved documents to it, and Ⓓ's address was recorded into the 'from' field of the same record as query 'Wireless LAN' in Ⓐ's Q/RDH. Since the documents include a content created by Ⓔ's user, Ⓔ's address is seen in the 'original' field of the content.

After getting these histories, if Ⓐ sends another query which is similar to 'Wireless LAN', say 'LAN', Ⓐ not only can find Ⓓ in a 'from' field of Q/RDH, but also find Ⓔ from an 'original' field of the content file by calculating a similarity between the query and the content file. Accordingly Ⓐ sends the query to both Ⓓ and Ⓔ.

The figure also shows that Ⓒ received query 'Mini Note PC' from Ⓑ, and both the query and Ⓑ's address were recorded into the Q/SAH. Even if Ⓒ has not sent a query, it can find information related to the queries it received using its Q/SAH. Therefore when Ⓒ sends a query, say 'Note PC', it will find Ⓐ and Ⓑ with the Q/SAH and can consequently send the query to them.

2.4   The Effect of Two Histories: Q/RDH and Q/SAH

As mentioned in the previous section, both Q/RDH and Q/SAH help to find target agents to send a query to. If an IR agent can find a sufficient number of target agents, no 'query multicasting' is carried out. Both histories, consequently, help to reduce communication loads between agents.

As a user creates more information, his/her IR agent can return the retrieved results to more queries. Such an IR agent consequently receives more queries from other agents. Thus, the agent accumulates more information sources comprised of pairs of a query and a sender agent's address in its Q/SAH. As the results, more queries the agent which has rich information receives, more information sources it acquires. That leads to the emergence of a 'give and take' effect.

Further, the user's positive or negative judgments concerning the retrieved results could be embedded into them in Q/RDH. These user evaluations are expected to be useful for finding target agents which will return relevant information, creating a collaborative filtering effect. This will be pursued in future work.

3   Experiments

3.1   Implementation with KODAMA

The ACP2P method was implemented with Multi-Agent Kodama (Kyushu university Open & Distributed Autonomous Multi-Agent) [13]. Kodama comprises hierarchical structured agent communities based on a portal-agent model. A portal agent(PA) is the representative of all member agents in a community and allows the community to be treated as one normal agent outside the community. A PA has its role limited in a community, and the PA itself may be managed by another high-level portal agent. A PA manages all member agents in its community and can multicast a message to them. Any member agent in a community can ask the PA to multicast its message.

All agents form a logical world which is completely separated from the physical world consisting of agent host machines. That means agents are not network-aware, but are organized and located by their places in the logical world. This model is realized with the agent middle-ware called Agent Communication Zone(ACZ for short). ACZ is primarily designed to act as a bridge between distributed physical networks, creating an agent-friendly communication infrastructure on which agents can be organized in a hierarchical fashion more easily and freely. ACZ is also designed to realize a peer-to-peer communication between agents.

A Kodama agent consists of a kernel unit and an application unit. The kernel unit comprises the common basic modules shared by all Kodama agents, such as the community contactor or message interpreter. The application unit comprises a set of plug-in modules, each of which is used for describing and realizing a specialized or original function of agents. All agents of ACP2P are realized by implementing their functions in plug-in modules of Kodama's application unit.

3.2   Preliminaries

We used the Web pages of Yahoo! JAPAN[14] for the experiments. The Web pages used are broadly divided into five categories: animals, sports, computers, medicine, and finance. Each of them consists of 20 smaller categories, which are selected in descending order of the number of Web pages recorded in a category. An IR agent is assigned to each selected category, and thus 100 IR agents are created and activated in the experiments. A category name is used as the name of an IR agent, and the Web pages in the category are used as the original documents of the agent as described in section 2.2. All IR agents were assigned to the same one community for simplicity.

We conducted experiments to show how two histories help to reduce communication loads between agents looking for information relevant to a query, and how Q/SAH helps in searching for new information sources. To perform the experiments, we compared three methods : 1) ACP2P with a Q/SAH(wQ/SAH for short), 2)ACP2P without a Q/SAH(woQ/SAH for short), and 3) Simple method always employing a 'multicast' technique (MulCST for short).

In the experiments, two query sets:QL=1 and QL=2, were used. QL=1 and QL=2 consist of 10 queries, whose query length is one and two, respectively, where query length means the number of terms in a query. When using queries belonging to QL=1, 10 nouns are extracted from every category assigned to each IR agent in descending order of their frequency of occurrence in the category. Each of noun is used as a query of the IR agent. When using those belonging to QL=2, 5 nouns are extracted and the combinations of the extracted 5 nouns taken in pairs create 10 queries.

3.3 Similarity Measure for Retrieving Information Relevant to a Query

In order to find the requested number of target agents to be sent a query, we calculated $Score(query, t\_agent)$, which returned the similarity value between query $query$ and target agent t_agent, with equation (1); $Score(query, t\_agent)$ becomes higher if t_agent sends a greater number of similar queries and returns more documents related to $query$.

$$Score(query, t\_agent) = \sum_{i=1}^{k} \cos(\boldsymbol{query}, \boldsymbol{qh_{d_i}})$$

$$+ \sum_{i=1}^{m} (\cos(\boldsymbol{query}, \boldsymbol{qh_{sa_i}}) + \varphi(i)) + \sum_{i=1}^{n} Sim_d(\boldsymbol{query}, \boldsymbol{doc_i}) \quad (1)$$

$$\varphi(i) = \begin{cases} \delta \text{ if } qh_{sa_i} \text{ is the query sent by other IR} \\ \quad \text{agent directly.} \\ 0 \text{ otherwise} \end{cases}$$

In equation (1), $query$ consists of $w_1, ..., w_m$, and w_i $(1 \leq i \leq m)$ is a term in $query$. qh_d and qh_{sa} represent a query in a record of Q/RDH and Q/SAH, respectively. The first term $\sum_{i=1}^{k} \cos(\boldsymbol{query}, \boldsymbol{qh_{d_i}})$ returns the total score of the similarities between $query$ and each of k number of queries sent to t_agent. The second term $\sum_{i=1}^{m} (\cos(\boldsymbol{query}, \boldsymbol{qh_{sa_i}}) + \varphi(i))$ represents the score between *query* and qh_{sa_i}, which is the i_th of m queries sent by t_agent in Q/SAH. $\varphi(i)$ is a weight to consider the importance of 'direct sending of a query'. If qh_{sa_i} is sent directly by t_agent, δ is added to the score. In the experiment, we set it to 0.1 from our empirical experience. The last term $\sum_{i=1}^{n} Sim_d(\boldsymbol{query}, \boldsymbol{doc_i})$ is the total score of similarities between *query* and each of n documents originally created or just owned by the user of t_agent. $Sim_d(query, doc)$ represents the similarity between *query* and the content of retrieved document *doc*. It is calculated with the following equation, which is a simplified of BM15[15].

$$Sim_d(query, doc) = \sum_{i=1}^{m} \frac{tf_i}{tf_i + 1}$$

Where tf_i represents the frequency of occurrence of w_i in *doc*.

After calculating $Score(query, t\_agent)$ for each IR agent t_agent in the Content file and two histories : Q/RDH and Q/SAH, the requested number (RN) of target agents will be selected in the descending order of $Score(query, t\_agent)$, which value should be more than 0. Whenever the RN of agents is not found, the 'query multi-casting' technique will be employed by a portal agent. At that time, all answers will be returned to the portal agent. If a target IR agent finds information relevant to *query*, it returns a 'YES' message, otherwise a 'NO' message. The judgment as to whether or not a document is relevant to a query is made according to the criterion of Boolean AND matching, that is, if the document includes the conjunctions of all terms in *query*, it will be judged relevant, otherwise irrelevant.

3.4   Experimental Results

First, we conducted the experiment to show how much ACP2P with Q/SAH worked for reducing communication loads. To do that, we investigated the change of the average number of messages exchanged by each IR agent for every query input. The experiments were conducted with two query sets: QL=1 and QL=2 on which tests with 4 different requested numbers: RN=3, 5, 7 and 10. In both query sets' cases, the average number of messages exchanged by each IR agent is reduced for every query input. Due to the limitation of the space, we will show it elsewhere[16].

Next, for both QL=1 and QL=2, we compared the three methods: wQ/SAH, woQ/SAH and MulCST. The RN was set to 10. The results are shown in Fig. 5. In both cases, the number of exchanged messages in MulCST almost did not change for every query input, while that for both wQ/SAH and woQ/SAH was reduced. In addition, wQ/SAH had better performance than woQ/SAH. That means Q/SAH worked well to look up relevant information with less communication efforts and made the positive feedback effect.

We also compared three methods for the average number of documents acquired by each IR agent. The results are shown in table 4. Except for the case of RN=3 of QL=2, there was little difference between wQ/SAH and MulCST.

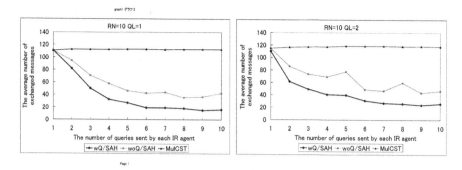

Fig. 5. The average number of messages exchanged by each IR agent for every query input, where QL=1 is the left and QL=2 the right. RN=10 in both cases.

Table 4. Comparison on average number of acquired documents when query length is 1 (left) and 2 (right)

QL=1, RN=	3	5	7	10	QL=2, RN=	3	5	7	10
wQ/SAH	269.1	385.3	443.0	497.6	wQ/SAH	54.9	126.3	178.9	226.8
woQ/SAH	258.8	331.6	424.9	476.4	woQ/SAH	54.8	96.8	150.0	208.3
MulCST	233.8	366.4	421.3	487.0	MulCST	85.3	148.4	191.0	232.6

4 Related Work

There is lots of work related to the topics touched in this paper, such as distributed information retrieval(DIR), P2P file searching, collaborative filtering and so forth. DIR selects some IR systems to send a query, aggregates the results returned by the selected IR systems, and presents them to a user. Before selecting the IR systems to be sent a query, the resource description of each IR system is often created[1]. In the ACP2P method, Q/RDH incrementally creates an effect similar to the resource description, and furthermore, Q/SAH works as good heuristic in finding relevant information.

A lot of P2P file searching systems such as Freenet[10], Chord[9], Gnutella[11] and Napster[12] have been proposed. Freenet and Chord are carried out in a pure P2P computing architecture. They neither employ 'broadcast' techniques like Gnutella, nor have a centralized server machine like Napster. Freenet provides information-sharing and information-finding functions among anonymously distributed nodes. Although Chord does not provide anonymity of nodes, it has an efficient protocol for looking up nodes. Their node searching strategies are conducted according to keywords attached to the information of the nodes. On the other hand, The ACP2P method makes use of the content information of documents, and two histories: Q/RDH and Q/SAH to search for target agents with relevant information. In particular, Q/SAH provides similar effects to link analysis like PageRank[17] or HITs algorithm[18] and makes a natural collaborative filtering effect emerge.

I-Gaia[19] is an application layer for information processing in the DIET architecture, which is a Multi-Agent System development platform. ACP2P is also a Multi-Agent-based application, but it does not use a mediator agent like t-infocytes of I-Gaia to learn appropriate paths between agents in sending queries or publishing documents.

Lots of work on the field of Collaborative Filtering (e.g.[5],[6],[20],[21],[8]) has been done. Most of it however assumes the server-client computational model and needs a procedure to collect all data from other nodes explicitly. The ACP2P method takes a distributed data management method with agent communities based on a P2P computing architecture, and makes a natural collaborative filtering effect emerge, with two histories.

5 Conclusion and Future Work

We discussed an agent-community-based peer-to-peer information retrieval method, called the ACP2P method, which used the content of retrieved document files and two histories: Q/RDH and Q/SAH to find target agents to be sent a query. The method was implemented with Multi-Agent System Kodama.

We conducted several experiments to show whether or not two histories helped to reduce communication loads between agents in searching for information relevant to a query, and whether or not Q/SAH helped in looking up new information sources. The experimental results showed the efficiency of the ACP2P method and the usefulness of two histories for looking up new information source. We also investigated and confirmed that the number of agents exchanging query messages together was increased by Q/SAH although we could not describe the detail about it due to the limitation of the space.

We are currently investigating the accuracy of or a method ranking retrieved results, and considering how we can make use of user feedback embedded into the results. Developing an effective method for creating hierarchical agent communities to allocate agents to at the initial stage and mining two histories for catching a change of user interests are future work.

Acknowledgment

This research was partly supported by the Telecommunication Advancement Organization(TAO) of Japan, under the grant to "the Research on Management of Security Policies in Mutual Connection" and by the Japan Society for the Promotion of Science under the Grant-in-Aid for Scientific Research (C) No. 16500082.

References

1. Callan, J., Connell, M.: Query-based sampling of text databases. ACM Transactions on Information Systems **19** (2001) 97–130
2. Lang, K.: NewsWeeder: learning to filter netnews. In: Proceedings of the 12th International Conference on Machine Learning, Morgan Kaufmann publishers Inc.: San Mateo, CA, USA (1995) 331–339

3. Schafer, J.B., Konstan, J.A., Riedi, J.: Recommender systems in e-commerce. In: Proceedings of the 1st ACM Conference on Electronic Commerce. (1999) 158–166
4. Yimam-Seid, D., Kobsa, A.: Expert finding systems for organizations: Problem and domain analysis and the demoir approach. Journal of Organizational Computing and Electronic Commerce **13** (2003) 1–24
5. Goldberg, D., Nichols, D., Oki, B.M., Terry, D.: Using collaborative filtering to weave an information tapestry. Communications of the ACM **35** (1992) 61–70
6. Resnick, P., Iacovou, N., Suchak, M., Bergstrom, P., Riedl, J.: Grouplens: Open architecture for collaborative filtering of netnews. In: Conference on Computer Supported Cooperative Work. (1994) 175–186
7. Good, N., Schafer, J.B., Konstan, J.A., Borchers, A., Sarwar, B.M., Herlocker, J.L., Riedl, J.: Combining collaborative filtering with personal agents for better recommendations. In: AAAI/IAAI. (1999) 439–446
8. Sarwar, B., Karypis, G., Konstan, J., Riedl, J.: Item-based collaborative filtering recommendation algorithms. In: WWW10. (2001) 285–295
9. Stoica, I., Morris, R., Karger, D., Kaashoek, M.F., Balakrishnan, H.: Chord: A scalable peer-to-peer lookup service for internet applications. In: Proceedings of the 2001 conference on applications, technologies, architectures, and protocols for computer communications. (2001) 149–160
10. Clarke, I., Sandberg, O., Wiley, B., Hong, T.W.: Freenet: A distributed anonymous information storage and retrieval system. Designing Privacy Enhancing Technologies: International Workshop on Design Issues in Anonymity and Unobservability, http://www.doc.ic.ac.uk/~twh1/academic/ (2001)
11. Gnutella: http://gnutella.wego.com/ (2000)
12. Napster: http://www.napster.com/ (2000)
13. Zhong, G., Amamiya, S., Takahashi, K., Mine, T., Amamiya, M.: The design and application of kodama system. IEICE Transactions INF.& SYST. **E85-D** (2002) 637–646
14. Yahoo: http://www.yahoo.co.jp/ (2003)
15. Robertson, S.E., Walker, S.: Some simple effective approximations to the 2-poisson model for probabilistic weighted retrieval. In: Proceedings of the 17 Annual International ACM SIGIR Conference on Research and Development in Information Retrieval. (1994)
16. Mine, T., Matsuno, D., Kogo, A., Amamiya, M.: Design and implementation of agent community based peer-to-peer information retrieval method. In: CIA 2004, LNAI 3191. (2004) 31–46
17. Brin, S., Page, L.: The Anatomy of a Large-Scale Hypertextual Web Search Engine. In: Proc. of 7th International World Wide Web Conference:WWW7 Conference. (1998)
18. Kleinberg, J.M.: Authoritative sources in a hyperlinked environment. Journal of the ACM **46** (1999) 604–632
19. Gallardo-Antolin, A., Navia-Vasquez, A., Molina-Bulla, H.Y., Rodriguez-Gonzalez, A.B., Valverde-Albacete, F.J., Cid-Sueiero, J., Figuieras-Vidal, A., Koutris, T., Xiruhaki, C., Koubarakis, M.: I-Gaia : an Information Processing Layer for the DIET Platform. In: the first international joint conference on Autonomous Agents and Multi Agent Systems (AAMAS). (2002) 1272–1279
20. Herlocker, J.L., Konstan, J.A., Borchers, A., Riedl, J.: An algorithmic framework for performing collaborative filtering. In: SIGIR99. (1999) 230–237
21. Melville, P., Mooney, R.J., Nagarajan, R.: Content-boosted collaborative filtering. In: SIGIR-2001 Workshop on Recommender Systems. (2001)

Emergent Structures of Social Exchange in Socio-cognitive Grids

Daniel Ramirez-Cano and Jeremy Pitt

Intelligent Systems & Networks Group, Dept. of Electrical & Electronic Eng.,
Imperial College London, SW7 2BT, UK
{d.ramirez, j.pitt}@imperial.ac.uk

Abstract. Several different forms of peer-to-peer interactions, associations and interpersonal relations between human and artificial intelligences are described. We build upon a new form of grid computing which integrates human and artificial 'processes' in electronically saturated physical spaces, called *socio-cognitive grids*. We start from the analysis of three scenarios in P2P applications: *digital rights management, mass user support* and *customer-to-customer interaction*. These enable us to identify those factors that motivate the computing components in the socio-cognitive grids to form social structures, individually incorporating socio-cognitive intelligence and social awareness. In order to study the emergent properties of these social structures, such as reciprocity, social exchange and social networking, we need a theory that will help us understand the dynamics of social integration and support. We explore the use of a classical sociological theory of social structures and interpersonal relations. Subsequently we outline the components of a software simulation built on this theory and designed to formalize and evaluate this socio-computational intelligence. Ultimately our main aim is to analyse and understand those emergent properties that lead to the formation of stable and scalable social structures in socio-cognitive grids.

1 Introduction

The widely-recognized peer-to-peer (P2P) applications (e.g. Napster, Kazaa, E-mule) have gained their reputation first due to the large number of people using these file-sharing applications and second because of the public controversy that has been created about whether or not their use is legal. However they have proved to have a similar or even greater impact in the research community since they have demonstrated that from basic peer-to-peer interactions it is possible to dynamically create social networks within which people can collaborate by sharing and retrieving information [1].

Along similar lines, grid computing has shown that the concept of sharing distributed resources is a feasible solution to large-scale applications that are characterized by resource-intensive processes [2]. Grid computing demonstrates that researchers in different geographical locations are willing to collaborate by sharing resources and working together to solve complex problems that require

G. Moro, S. Bergamaschi, and K. Aberer (Eds.): AP2PC 2004, LNAI 3601, pp. 74–85, 2005.

large-scale data analysis, terabytes of storage space and expert collaboration. Some of the programmes in the exploration of grid computing include EuroGrid and Grid Interoperability (GRIP) [3] in the European Union and the Globus alliance in the U.S. [4].

On the other hand both the EU i3 research programme [5] and the field of pervasive (or ubiquitous) computing recognise and promote the idea that physical spaces can be transformed and treated as electronic environments. Such programmes focus on the software and hardware technologies required to create environments which are saturated with computing devices and wireless communications, yet appear to be gracefully integrated to the human user(s), and indeed, gracefully integrate the human users themselves.

Thus the label "socio-cognitive grid", as identified in [6], stands for the term which embraces the fields of grid, pervasive and P2P computing. This includes applications which are composed of electronically saturated physical spaces together with seamlessly integrated and interacting intelligences, both computational processes (e.g. software agents) and human processors (i.e. people). Therefore, a socio-cognitive grid is an extension and generalisation of grid computing, whereby if grid computing is defined as applying resources from many networked computers – at the same time – to a single problem (see e.g. [7]), then socio-cognitive grid computing can be defined as the application of resources from many networked computers and people at the same time to the same single problem. Note then that the kinds of "problem" to be solved are not just those addressed by grid computing, (i.e. large-scale data processing on networked computers for problems such as mapping the human genome), but can be far more open-ended and non-directed, with radically new forms of human-computer interaction and computer-mediated human-human interaction.

In this paper we present the analysis of different forms of peer-to-peer interactions, associations and interpersonal relations that lead to the formation of social structures in these socio-cognitive grids. Our aim is to characterize those emergent properties of the social structures that contribute to stable and scalable socio-cognitive grids. Furthermore, we are particularly interested in understanding how individual behaviour of peers affects social cohesion, social support, collaboration and self-regulation in the socio-cognitive grids.

We start with the description of three scenarios (Section 2) in P2P applications: *digital rights management (DRM), mass user support (MUS)* and *customer-to-customer interactions (C2C)*. In Section 3 we present the analysis of these three scenarios which, we argue, will disclose the essential properties of the social associations in the socio-cognitive grids, while in Section 4 we adopt and adapt the classical sociological theory of interpersonal relationships and social structures, in an attempt to define the associations occurring inside the socio-cognitive grids. In Section 5 we determine the basic building blocks needed for the construction of a software simulation to evaluate the concepts defined previously. Finally Section 6 closes the analysis with a discussion and suggests further work towards the logical formalization of social structures in socio-cognitive grids.

2   Scenarios

In this section we present P2P applications in which:

- information trading is involved [8], i.e. information itself is a commodity, and the exchange of content is the main form of interaction in the society.
- there is some subjective utility to the end user consuming the content, for example, when asking a question, both the quality of the answer and reliability of the source are factors to be considered.
- there is a community of content creators and end users which thrives on mutual collaboration, i.e. the only rule-enforcing authority is self-regulation in the form of reputation and social attitudes.
- some mechanism exists for evaluation and feedback about the quality of content and of behaviour to the community, in particular, anonymous behaviour is not tolerated and all actions have (social) consequences;
- given a community of users interacting via their agents, those agents themselves form an *open agent society* [8], and the interactions of the agents have corresponding consequences for the users in the human community.

2.1   Digital Rights Management

John is walking down the street and on his way he passes by a record shop. The shop has recently invested in a new software agent application, and a W-LAN and Bluetooth infrastructure which allows an agent to broadcast twenty-second samples of selected songs to everyone within proximity. An agent on John's mobile registers interest in a selection of the advertised songs and consequently accepts a sample of the latest song from a new band. After listening to the sample John wants to have the complete track, so he asks his personal agent to buy it. The transaction is made based on a digital rights system called LWDRM [9]. This technology allows John's agent to receive the song and a digital licence which is attached to the song. In return the agent in the record shop receives the payment through a micro-payments infrastructure (MPI) that charges the cost of the song directly to John's mobile phone bill. The digital licence that John bought authorizes him to share the song and since John rated it so highly, his agent decides to share it with John's friends which initiates an excellent word-of-mouth promotion for the new band.

2.2   Mass User Support

An important characteristic of socio-cognitive grids is the opportunity for social collaboration and coordination of people and their agents. In line with socio-cognitive grids, mass user support stands for the social coordination mechanisms and services that support a collective user-base [10]. Richard is looking for a flat in Liverpool and finally he finds the place that seems right to him. However Richard is new in town and he has no idea about the neighbourhood and the security of the area. Consequently he programs his agent to ask other agents in the vicinity and gather as much information as possible about the neighbour-hood and if possible about the flat itself while he visits the area for the second time. Interestingly, Richard's agent receives only good comments about the area but opportunely, after going through several mediators, his agent finds the agent of the previous tenant, George. According to George's agent the flat is not in satisfactory condition and even worse, has a nuisance problem coming from adjacent neighbours who have a band. Richard decides not to take the flat and keeps looking for more options.

2.3 Customer-to-Customer Interaction

Lucy is in her second year in college and she needs to find accommodation. Her personal agent starts interacting with her friends' agents in college asking for available accommodation close to the campus or at least information as to how to find accommodation. Michelle's agent receives Lucy's request and knows that Lucy is a very good friend who has helped Michelle several times in the past thus the agent decides to help Lucy. Michelle's agent knows that Rita (another student) is advertising an available room in her shared house thus extends Lucy's request to Rita. Lucy's and Rita's agents exchange information through Michelle's agent and Lucy's agent discovers that the room is within her price range, it is within 10 min. walking distance of campus and is available now. Rita's agent finds out that Lucy is a non-smoker and she is studying the same course as Rita. It is just what both girls are looking for. Lucy receives the information about Rita's house and five more options that her personal agent has found from people around, plus a list of suggestions about renting private housing that the agent retrieved from the private housing office. Lucy decides to consider Rita's offer therefore her agent contacts Rita's agent and the girls personally meet in the college's common room.

3 Requirements for the Realization of Socio-cognitive Grids

From the analysis of the scenarios described above we observe a set of techno-logical specifications that define the communication at the physical level and a different group of sociological specifications at the logical level that explain the social interactions and associations in the socio-cognitive grids.

3.1 Technological Specifications

We are interested in communities of people in which each person is portrayed by a software agent placed in a portable electronic device(e.g. PDA, mobile phone, laptop) which is enabled with a wireless transmission technology (e.g. Bluetooth, IEEE 802.11). Presumably these specifications are relatively straightforward to meet and technologically feasible since it is possible to build these systems with today's technology.

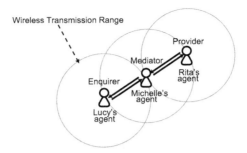

Fig. 1. Transient chains of communication

The agents communicate and interact but most importantly, share knowledge of the environment and exchange content. At the first level an agent is restricted by its transmission range to peer-to-peer interactions only with nearby agents. However we propose that this limitation can be overcome through transient chains of communication and agents behaving as content *providers* and message *mediators*. From the C2C scenario we can observe that Rita's agent is able to provide information to Lucy's agent, who is completely outside its transmission range, thanks to Michelle's agent intervention as mediator(see Fig. 1). Due to the concept of transient chains of communication the formation of social structures is not constrained by physical boundaries.

The arrangement shown in Fig. 1 is a representative picture of a community at a given instant. In reality we are dealing with a society in motion, ruled by the same principles as an ad hoc network topology [11]. The connections are dynamic, self-organizing and without any central authority. The nondeterministic movement of the peers causes unfinished interactions and exchanges which should have no influence on the stability of the society as far as the agents are aware of this fact. However this movement facilitates unplanned situations which create new opportunities.

3.2   Sociological Specifications

Socio-cognitive grids are possible due to the seamless integration and interactions of human and artificial intelligences, namely people and software (agents). From the scenarios we observe that these two types of intelligences interact in different forms defining a typology of interpersonal connections: people-to-software, software-to-software and people-to-people via software, as show in Fig. 2.

Let P be the domain of people, S be the domain of agents and E be the joint domain of people and agents, we can then characterise the types of interpersonal connections as follows:

$$\text{Interactions}: \quad X \quad \textbf{says to} \quad\quad Y, \quad\quad \text{where } X, Y: E$$

$$\text{Associations}: X \& Y \;\textbf{belong to} \; \textit{Community } \Theta, \text{where } X, Y: S$$

$$\text{Interpersonal relations}: \quad X \quad \textbf{is friend of} \quad\quad Y, \quad\quad \text{where } X, Y: S$$

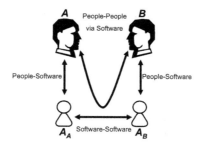

Fig. 2. Typology of interpersonal connections

The combinations of these forms of connection characterize the different applications. In Table 1 we show an instance of this typology in which we can see how it differentiates our scenarios. In DRM, the economic transactions can be completely handled by the interaction of agents. Moreover the agents are able to extend the friendship between users by giving recommendations to agents that belong to the same community. C2C implies traditional people-to-people interactions, however these interactions are extended via software agents which increase the probability to match buyers with sellers as a result of maximizing the amount of possible interactions. MUS is an all-inclusive application. Its essence lies in the association of people with the desire to share something they know is useful. The software is simply the medium to cooperate and create interpersonal relationships.

From the inspection of the interpersonal relations between people and software in the scenarios, we observe that social associations arise between members of the socio-cognitive grids. We presume that as the associations grow they will transform into complex social structures and interpersonal relationships. In order

Table 1. Typology applied to scenarios

	DRM	MUS	C2C
People-Software	John's agent selected a sample track for him and then John instructed his agent to buy the song	Richard received the critique of the flat from George's agent	Lucy instructed her agent to ask her friends for available accommodation
Software-Software	John's agent bought the song from the record shop agent	Richard's agent found George's agent through the agent mediators	Michelle's agent introduced Lucy's to Rita's agent
People-People via software	John recommended the purchased song to his friends, by way of their respective agents	George helped Richard as a result of the review that George programmed on his agent	Lucy met Rita thanks to Michelle's agent

to examine the emergent properties of the socio-cognitive grids we need a theory that will help us understand the dynamics of social integration and support in social structures.

4   Theory of Social Structures

In order to investigate the principles and motivations that lead to the formation of stable and scalable social structures in the socio-cognitive grids, we borrow the concept of social structure specified by classic exchange theorists from the field of sociology and defined in [12] as: "... a configuration of social relations among actors (both individual and corporate), where the relations involve the exchange of valued items (which can be material, informational, symbolic, etc)." This exchange of valued items is referred to as *social exchange* and is described by Blau [13] as "...voluntary *actions* of individuals that are motivated by the *returns* they are expected to bring and typically do in fact bring from others."

We adapt the concepts mentioned above to build a conceptualization of mutual exchange of digital content and support in the socio-cognitive grids, based on the anthropomorphism of social exchange between people-agent and agent-agent. The purpose of such anthropomorphism is not to replicate the complex behaviour of human social relations or to create simulations of human societies but to structure a formal framework capable of understanding how individual behaviour, decisions and interests affect the dynamics of the construction of social structures. We now describe those individual elements that, we suggest, generate preferable societies in which intelligences want to stay and to which others desire to belong.

4.1   Reciprocity in Social Exchange

According to Blau's notion of social exchange once someone receives a favour, assistance, information or any kind of service, he or she is implicitly obligated to return the favour. We consider that the concept of obligation is not necessarily (although it might be) enforced by legal bindings or a written contract but it is a voluntary return encouraged by the interest to:

- increase the probability of finding someone who is willing to help you when needed
- acquire reputation and recognition by the community
- contribute to the operation of efficient networks of social support

The dynamics of the society dictate reciprocity. An agent who gives something expects to receive something in return. If the received content is of high value then the agent will seek to give something again in order to receive more of that high valued content. On the other hand an agent who gives something but receives nothing in return or low-valued content will therefore be disappointed and will lose the motivation for any further reciprocal interaction.

At this point it is important to make the same clarification proposed by Blau between economic exchange and social exchange. An economic exchange is the formal exchange of content which is priced in advance at a specific value and thus an agreed value of the return, namely payment, is expected. Economic exchange does not entail reciprocity or voluntary returns. On the other hand social exchange suggests the bestowing of content with the promise of receiving something in return. In our study we assume socio-cognitive grids with the ability to support the existence of both economic and social exchange. e.g. DRM implies an economic exchange of digital data plus the appropriate digital rights for an economic remuneration. However, as we have describe in the DRM scenario, even in this kind of application both parties can benefit from an underlying social exchange of recommendations, reviews and opinions.

4.2   Social Networking

So far we have analysed the dynamics of the socio-cognitive grids from the perspective of an individual, from the inside to the outside. Now we move towards an external perspective to examine the formation and dissolution of societies, agreement and dissension and the general characteristics of networks of social exchange. We adopt Cook and Emerson's conceptualization of an exchange network which is defined [14] "as consisting of (1) a set of actors (either natural persons or corporate groups), (2) a distribution of valued resources among those actors, (3) for each actor a set of exchange opportunities with other actors in the network, (4) a set of historically developed and utilized exchange opportunities called exchange relations into a single network structure". From this definition it is possible to examine exchange networks in agent societies from the following perspectives:

- The creation of roles and the distribution of power understood as the influence that one agent exerts on the society and the degree to which others depend on it
- The distribution of the resources and the position of the agents in the network
- The association of agents sharing common believes and the consequent isolation from the society or the implementation of strict admission policies
- The factors that influence the formation, perpetuation, disintegration and isolation of social networks

However we believe the above perspectives are just an initial instance of a considerable range of approaches to the study of networks of social exchange in socio-cognitive grids.

5   Evaluation

We have proposed the adaptation of the classical theory of social structures based on social exchange as a framework to understand social interactions and associations in socio-cognitive grids. To evaluate the concepts discussed in Sections 3 and 4 we believe it is necessary to build a software simulation of the scenarios previously discussed. In this section we present the building blocks of such a simulator: data structures for social memory, an algorithm for updating social memory and a representation of the flow of content.

5.1   Social Memory

We have discussed how important it is that an individual remembers its actions towards every person in the society and the rewards obtained in return for its actions. The duality actions-returns can be used to build an assessment about the type and quality of each specific relationship. Furthermore, in order to transpose these notions to an artificial society of agents, we suggest to design agents with a structure that will allow them to remember all the interactions or exchanges with other members of the society. We present two structures based on the actions and returns described by Blau and on the roles of mediator and provider introduced in Section 3.1:

provider(Name, PP, MPP, Loc_p)	*mediator(Name, MP, MMP, Loc_m, TS)*
Name: Name of the agent	Name: Name of the agent
which provided the content	which mediated the content
PP: Provider Points	MP: Mediator Points
MPP: Mirror Provider Points	MMP: Mirror Mediator Points
Loc_p: List of content from providers	Loc_m:List of content from mediators
	TS: Time Stamp

Loc_p=content(Date, Subject, Rating)	*Loc_m=content(Date, Subject)*
Date: Date of the interaction	Date: Date of the interaction
Subject: Description of the content	Subject: Description of the content
Rating: Rating given to the content	

The structures are very similar and they are both built upon the same principle of social exchange. However, as will be described in Section 5.2, the algorithms and criteria to update the provider and mediator points are different. In both, the field *Name* records the identity of the interacting agent. The field *PP* reflects the opinion that the agent has about another agent, based on the quality of the content directly received from that agent (returns). Similarly the field *MP* shows an assessment of other agents but not as direct provider, instead functioning as a mediator through which the agent receives the content. The fields *MPP* and *MMP* are estimated in the same way as *PP* and *MP* respectively but they are a self-assessment of the behaviour (as provider or mediator) towards other agents (actions). The structure for mediators has an extra field labelled *TS* which contains a time stamp of the last interaction with every agent. As will be discussed in Section 5.2, *TS* is used as a time reference to calculate the value of *MP* and *MMP*. The *provider* and *mediator* structures give every agent a social memory which can be used to evaluate the relationship with other members of the community and make decisions about the level of cooperation with each individual.

Based on the principles of social networking previously discussed in Section 4.2, it is also convenient to equip the agents with the ability to remember the set of historically developed exchange opportunities. The fields *Loc_p* and *Loc_m* allow for a *List of Concepts* that contains the type and description (*Subject*) of the historically exchanged content; and the *Date* of the interaction. Additionally, for the list of providers, the evaluation of the content is included in the form of a *Rating* which is needed to calculate the value of *PP* and *MPP*. In turn these values are used as the main parameters for the selection criterion to choose recipients.

5.2 Updating Social Memory

The values of *PP*, *MP*, *MPP* and *MMP* are dynamically updated with every applicable interaction. They are always a number in the range between 0 and 1 anticipating a future exploration of probabilistic models in the algorithms for updating social memory and flow of content.

The process to update the value of *PP* and *MPP* is the weighted sum $NewValue = \omega_1 h_1 + \omega_2 h_2$. In the equation, h_1 is the same evaluation of content or *Rating* introduced in Section 5.1. h_2 is called the *Friendship level* and it is a method to express the appreciation towards the interacting agent. The supporting premise states that it is difficult to prove to be friend, thus new relations are slow to build even if the exchanged content is good. However once a threshold is reached and therefore the status of friend acquired then errors are easily forgiven and favours highly appreciated. The friendship level *(y)* is defined as an exponential transformation given by the equation $y = Ce^{kx} - C$ where x is the old value of *PP* or *MPP* before the new interaction; C and k are constants defined according to friendship thresholds in the user's profile. Finally ω_1 and ω_2 are weights assigned to the rating and friendship level and vary according to the agent implementation and user's profile.

On the other hand, the process of updating the value of *MP* and *MMP* is based on the assumption that usually a relationship deteriorates with time unless it is constantly nourished. This degradation is represented using a time decay function. The shape of the function is defined by the user profile. However it is suggested, as a first approach, that the desired time decay function follows the path of an exponential decay function which takes the difference between the current time and the time stamp *TS* as its main variable. On the contrary, every new interaction increases again the value of *MP* and *MMP*. For consistency, we suggest that the increase rate should be defined using the same principle and function used to calculate the previously defined *Friendship level*.

5.3 Flow of Content

From the scenarios of Section 2 we observe that while Richard wants to reach as many people as possible so that the probability of receiving specific information about the flat is higher, Lucy on the other hand is going to share a house therefore she wants to ask only people she trusts, namely her friends. On this basis we define two methods of selecting recipients: *Broadcast* and *Limited*. Broadcast is the propagation of the request to reach as many recipients as possible whereas Limited implies a specific selection criterion which Homans [15] suggests is directly related to the value of the exchanged content and the probability of obtaining it: "In choosing between alternative actions, a person will choose that one for which, as perceived by him at the time, the value (degree of reward) V, of the result, multiplied by the probability, p, of getting the result, is the greater". We propose that every agent is motivated to choose recipients by different priorities and the ensemble of all these possibilities characterises the profile of the society.

However it is possible to define the variables to be considered when making a selection. The main purpose of every selection is to maximize the benefit and the probability of obtaining it. Generalizing a simple interaction between two agents, namely A and B, the probability that an agent A will choose agent B as a recipient is determined by: (i) the quality of the relationship between them (defined by PP), (ii) the probability that agent B would have the answer (defined by Loc_p) and (iii) the probability agent B will provide agent A with the answer given that agent B has the answer (defined by a condition of reciprocity between PP and MPP).

On the other hand the probability that agent B will respond to agent A is also given by a condition of reciprocity between PP and MPP (or MP and MMP in the case where agent B does not have the answer but knows agent C who might have it). We anticipate that this condition of reciprocity will be similar to the well known cooperation strategy TIT FOR TAT [16]; but to find the conditions that lead to stable interactions and scalable societies is precisely the purpose of the simulation.

6   Discussion and Further Work

In this paper we have discussed the idea of interpersonal relations, social structures and cooperation within the framework of socio-cognitive grids. We borrowed concepts from the sociological classical theory of interpersonal relationships and social structures. We recognize that social exchange and social networking are the building blocks of stable socio-cognitive grids. We suggest that societies within the socio-cognitive grids will build social relations based on the exchange of valued items and the creation of social networks. Furthermore this social exchange should be self-regulated by social norms, such as reciprocity, that emerge from within the society. From this point on we are now interested in four main paths: (i) the evaluation of these propositions through the construction of a software simulation based on the outline given in Section 5, (ii) the logical formalization of the social relations that shape the social structures, (iii) a deeper exploration of the utilized sociological theories and the incorporation of more complex social relations and (iv) the incorporation of the ideas of digital blush, shame and embarrassment [17] as new mechanisms of self-regulation.

Acknowledgements

This work has been supported by the National Council for Science and Technology in Mexico (CONACYT) and the EPSRC project 'Theory and Technology of Norm-Governed Self-Organising Networks' (GR/S74911/01).

References

1. Yu, B., Singh, M.P.: Searching social networks. In: Proceedings of the Second International Joint Conference on Autonomous Agents and Multiagent Systems. (2003) 65 – 72
2. Berman, F., Fox, G., Hey, T.: Grid Computing: Making the Global Infrastructure a Reality. Wiley (2003)
3. EuroGrid and Grid Interoperability (GRIP). http://www.eurogrid.org (2004)
4. The Globus Alliance. http://www.globus.org (2004)
5. Intelligent Information Interfaces. http://www.i3net.org/ (2004)
6. de Bruijn, O., Stathis, K.: Socio-cognitive grids: The net as a universal human resource. In: Proceedings of Tales of the Dissapearing Computing. (2003)
7. Hey, T., Trefethen, A.: The UK e-Science core programme and the grid. Future Generation Computing Systems (FGCS) 18 (2002) 1017–1031
8. Pitt, J., Mamdani, A., Charlton, P.: The open agent society and its enemies: a position statement and research programme. Telematics and Informatics 18 (2001) 67–87
9. van der Pluijm, H.: Pay once, share often with LWDRM. http://www.wired.com/ news/digiwood/0,1412,62739,00.html (2004) Wired News.
10. Kurumatani, K.: Mass user support by social coordination among users. In Kurumatani, K., Chen, S., Ohuchi, A., eds.: Proceedings IJCAI-03 Workshop on Multiagent for Mass User Support. (2003) 58–59
11. Wu, J., Stojmenovic, I.: Ad hoc networks. IEEE Computer 18 (2004) 29–31
12. Cook, K.S., Whitmeyer, J.M.: Two approaches to social structure: Exchange theory and network analysis. Annual Review of Sociology 18 (1992) 109–127
13. Blau, P.M.: Exchange and Power in Social Life. John Wiley and Sons (1964)
14. Cook, K.S., Emerson, R.M., Gillmore, M.R.: The distribution of power in exchange networks: Theory and experimental results. The American Journal of Sociology 89 (1983) 275–305
15. Homans, G.C.: Social Behaviour. revised edn. Harcourt-Brace (1974)
16. Axelrod, R.: The Evolution of Cooperation. Basic Books (1984)
17. Pitt, J.: Digital blush: towards shame and embarrassment in multi-agent information trading applications. Cognition, Technology and Work 6 (2004) 23–36

Permission and Authorization in Policies for Virtual Communities of Agents

Guido Boella[1] and Leendert van der Torre[2]

[1] Dipartimento di Informatica - Università di Torino-Italy
guido@di.unito.it
[2] CWI Amsterdam and TU-Delft - The Netherlands
torre@cwi.nl

Abstract. We study the design of policies for virtual communities of agents based on peer-to-peer systems or the grid infrastructure. In a virtual community agents can play both the role of resource consumers and the role of resource providers. Moreover, the agents remain in control of their resources, and therefore we distinguish between the authorization to access a resource given by the virtual community and the permission to do so issued by the resource providers. We propose a logical multiagent framework for virtual communities that distinguishes *three* roles: resource consumption, provision, as well as authorization.

1 Introduction

Peer-to-peer systems and the grid infrastructure allow to create virtual communities. For example, Pearlman *et al.* [1] define a virtual community as a large, multi-institutional group of individuals using a set of rules, a policy, to specify how to share their resources, such as disk space, bandwidth, data, online services, *etc*. In order to control the distributed nature of peer-to-peer or grid systems, policies are defined using norms, i.e., deontic notions like obligations, prohibitions and permissions. Inspiration comes from, amongst others, computer security and distributed systems [2,3]. For example, in the Kazaa file sharing system a user is obliged by the system to share files, otherwise his bandwidth for downloading files is reduced as a sanction.

However, policies in virtual communities are more complex than policies in traditional distributed systems, due to the following reasons.

- Every agent in the community can play both the role of a resource consumer as well as that of a resource provider. Resource providers retain the control of their resources and they specify in local policies the conditions of use of their resources.
- When there is a central manager, it permits agents to access the resources which it owns and controls, according to the policies defined by itself. In contrast, in virtual communities, access control cannot be directly implemented, since nobody owns all the resources.
- Resource providers implement local access control according to the community's security policies. However, they should not be overburdened by the task of updating the policies as they change and new members join the community.
- Agents who participate to the community are heterogeneous and change frequently, so they cannot be assumed to be always cooperative and to stick to the system

G. Moro, S. Bergamaschi, and K. Aberer (Eds.): AP2PC 2004, LNAI 3601, pp. 86–97, 2005.
© Springer-Verlag Berlin Heidelberg 2005

policies, concerning both requesting access to resources and providing access to their resources.

The problem of designing policies for virtual communities has been raised recently, e.g., by Pearlman *et al.* [1] and Sadighi Firozabadi and Sergot [4]. Pearlman *et al.* argue that the solution is "to allow resource owners to grant access to blocks of resources to a community as a whole, and let the community itself manage fine-grained access control within that framework". The centralized management of resources owned by the single resource providers is performed by a *Community Authorization Service* (or CAS): "A community runs a CAS server to keep track of its membership and fine-grained access control policies. A user wishing to access community resources contacts the CAS server, which delegates rights to the user based on the request and the user's role within the community. These rights are in the form of capabilities which users can present at a resource to gain access on behalf of the community".

In this paper we discuss the design of virtual communities policies composed by prohibitions, permissions and authorizations. We address the following problems.

1. Is the task of *authorizing* requests performed by the CAS - henceforth called also authority - identical to the task performed by a resource provider when it *permits* access? How should permissions and authorizations be distinguished and how are they related? What is the relation between the CAS and the resource providers?
2. How can a resource provider delegate to the CAS the power of authorizing resource consumers and why can the power to issue permissions not be delegated?

We analyze these distinctions using our framework for normative multiagent systems [5]. As Jin and Liu [6] notice, multiagent systems "have been widely used in peer-to-peer computing" and "it is regarded as a perfect match to integrate peer-to-peer computing and agent-based systems, because since their inception, multiagent systems have been always thought of as network of peers". We extend the use of multiagent systems in peer-to-peer to normative multiagent systems, i.e., multiagent systems regulated by norms.

We use the following example in this paper, based on Pearlman *et al.*'s description of the process of accessing a resource in a virtual community. When a resource provider a_3 wants to join a community, it informs the CAS a_2, which replies with the requirements on how its resource must be shared with the community. When a resource consumer a_1 wants to access the resource of agent a_3, it must not only authenticate itself with agent a_2 providing its credentials, but it must also get a proof that its request conforms to the community's access policy. This proof is expressed by a *capability* (e.g., a X.509 certificate) provided by agent a_2 to a_1, which identifies the agent and states that it is *authorized* to access the resource. Now, agent a_1 can make the actual request to a_3, forwarding it the capability. After checking the truthfulness of the capability, agent a_3 replies to a_1. In a virtual community, agent a_3 maintains the control of its resource: the request is granted only if it is also permitted by the local policy of agent a_3. Hence, the authorization by agent a_2 contained in the capability is not enough for agent a_1's request being granted.

This paper is organized as follows. In Section 2 we discuss the notion of authorization. In Section 3 we introduce the formal agent model with the definition of norms (prohibitions and permissions), authorizations and delegation, which is illustrated by the above scenario. In Section 4 we discuss the theoretical foundations of the design and in Section 5 we summarize the results of the paper.

2   Authorization

A first cue that authorization and permission have different properties is found in the ordinary use of the terms authorization and permission. E.g., for the Cambridge Advanced Learner's Dictionary [7] permitting is "to allow something", "to make it possible for someone to do something, or to not prevent something from happening", while authorizing means "to give someone *official* permission to do something". Moreover, dictionaries of law like [8] argue that authorizations and permissions are related but different concepts, and that authorizations do not create new permissions.

In virtual communities, the authorization issued by the CAS is conceptually different from the permission granted by the resource provider, because the power of issuing permissions requires being in control of a resource. Resource providers delegate to the CAS the power to issue authorizations, but not the power to issue permissions. The notion of authorization to access a resource and the notion of permission should be kept distinct to correctly model of the situation. Moreover, they should be kept apart to prevent dangerous misunderstandings in the design of access policies.

Consider the following example. An agent a_3 joins some virtual community; it will both use the resources provided by the community, say downloading shared files, and provide its resource to the other members of the community, say some of its disk space to store files: agent a_3 plays both the role of a resource consumer, $c(a_3)$, and that of a resource provider, $p(a_3)$. Since agent $p(a_3)$ controls its disk space (it is the only one who can decide that storing or retrieving files take place), it regulated the access to the disk by means of some local policy: prohibitions and permissions. E.g., it prohibited to read files during the day and it prohibited to store files exceeding 2.5Mb.

When agent a_3 joins the community, it agrees that also some other members use its disk space resource. In principle, agent $p(a_3)$ could modify the policy regulating the access to its resource: e.g., by maintaining the prohibitions to read file during the day and to store large files, and by adding the permission about which members of the community can store and retrieve files on its disk space. However, this solution imposes an heavy burden. Even if the problem of authenticating which are the current users of the community can be dealt with by some trusted third party who gives them e-certificates, another problem remains: which members of the community are the ones which the community currently wants that they can access the resource and under which conditions they can do so. Moreover, the community's access policies may change with time, so that agent $p(a_3)$ should be kept informed and should modify the norms (prohibitions and permissions) regulating access to the resource it owns. The complexity of modifications could also introduce unwanted errors in the access policy of agent $p(a_3)$.

What is needed is a solution which transfers the burden of managing the community policies to other agents, playing the role of authorities, which have the knowledge and resources to perform this task. However, it is impossible to say that the CAS $u(a_2)$ changes the prohibitions and permissions posed by agent $p(a_3)$: in our model [5] norms are defined in terms of the goals which resource providers want to achieve concerning the use of their resources. The difficulty is that nobody can change the goals of an autonomous agent. Moreover, $u(a_2)$ is not in control of the resource so it cannot impose sanctions to motivate the respect of prohibitions. Finally, agent $p(a_3)$ wants to preserve its autonomy, so that it does not accept that someone else can change the prohibitions and permissions regulating access to its resource.

The solution is that agent $p(a_3)$ creates a permission saying that authorized agents can access the resource. But the decision to authorize agents to access the resource is delegated to the CAS $u(a_2)$ which has up to date knowledge on the system policies and members. Delegating the decision to authorize is easier than delegating permissions: the authorization is not a goal of the agent $p(a_3)$ but just a belief which can be induced by the CAS by issuing e-certificates and capabilities to the agents which are authorized. Moreover, it does not require that the delegated agent is in control of the resource.

When the set of agents which can be authorized changes as a consequence of new community policies, agent $p(a_3)$ does not have to change the prohibitions and permissions regulating access: new authorizations are created when the CAS $u(a_2)$ issues new capabilities (or, in our abstract terminology, $u(a_2)$ declares them authorized). The capabilities are recognized by agent $p(a_3)$ as the proof that the permission to access the resource applies to a consumer $c(a_1)$ requesting access to it.

Authorizations, thus, are the means used by authorities like the CAS to regulate the access of consumers to resources which they do not control. But there is no way to make authorized users access a resource without a permission by the resource provider which controls the resource: hence, authorizations are distinct from and presuppose permissions. An authorization is useless unless the resource provider permits authorized agents to access the resource it controls: authorizations change what is prohibited to an agent and legitimate an action but without introducing or removing any prohibition and permission.

Finally, nothing requires that agent $u(a_2)$, who is delegated the authority to authorize other agents, is itself permitted nor authorized nor delegated to authorize itself. The separation of institutional power from the permission to exercise it, identified by Makinson [9], is important for virtual communities. An organization could, e.g., outsource some administrative task such as assigning access rights to some agent without allowing it to have those access rights. In summary, the key notions are:

Prohibition is defined as a goal of resource providers. This is paraphrased as: Your wish (goal, desire) is my command (prohibition). The unfulfillment of the goal is considered as a violation and is sanctioned.

Permission is behavior which not considered by a provider as a violation and thus it is not sanctioned. The main role of permissions is to provide exceptions to prohibitions in a given context.

Authorization is a belief of a provider which appears as a condition in some permission it issued.

Declaration of authorization is an action of an authority which states that an agent can be considered authorized according to its own policy. Using Searle [10]'s terminology, in [5], we say that a declaration generates an actual authorization if it "counts as" an authorization for the resource provider.

Delegation establish who is considered as an authority. The declaration of someone
recognized as an authority turns into a belief of the resource provider that an agent
is authorized. A provider delegates the power to authorize to the CAS when it joins
the community.

3 A Formal Model

3.1 Individual Agent Design

In virtual communities there is no separation of resource providers from resource consumers, and they can play the role of authorities too. So we introduce a single set of
agents, which can each play one or more roles. For the individual agent design we are
inspired by the BOID architecture [11]. However, in contrast to the BOID architecture,
prohibitions are not taken as primitive concept. Beliefs, desires and goals are represented by conditional rules.

Definition 1 (Agents). *Let $\mathcal{A} = \{a_1, a_2, \ldots, a_n\}$ be a set of n agents. An agent $a_i \in \mathcal{A}$
can play three roles:*

1. *Resource consumer, denoted as $c(a_i)$: it can access resources to achieve its goals,
   is subject to norms regulating security, prohibitions and permissions, and also en-
   dowed with authorizations to access resources.*
2. *Resource provider, denoted as $p(a_i)$: it can provide access to the resources it owns.
   We call this the normative role, since it can issue norms, i.e., prohibitions and permissions about the access of a resource, and enforce their respect by means of
   sanctions, and delegate the power to authorize resource consumers.*
3. *Authority, denoted as $u(a_i)$: it can declare resource consumers authorized when
   they are requested to do so. They know that their declarations are considered as
   authorizations by the resource providers since they have been delegated the power
   to authorize resource consumers on behalf of resource providers.*

Actions, represented by *decision variables*, can have conditional and indirect effects
with a non-monotonic character. We assume that the base language contains boolean
variables and logical connectives. The variables are either *decision variables* of an
agent, which represent the agent's actions and whose truth value is directly determined
by it, or *parameters*, which describe the state of the world and whose truth value can
only be determined indirectly. Our terminology is borrowed from Lang *et al.* [12]. Institutional facts, a subset of the parameters, represent the legal classification of reality
made by agents.

Definition 2 (Decisions). *Let $A_i = \{m, m', m'', \ldots\}$, the decision variables of $a_i \in$
$\mathcal{A}$, and $P = \{p, p', p'', \ldots\}$, the parameters, be $n + 1$ disjoint sets of propositional
variables. Let the institutional facts I be a subset of P. A literal is a variable or its
negation. $d_i \subseteq A_i$ is a decision of agent a_i.*

The consequences of decisions are defined by the agent's epistemic state, i.e. its
beliefs about the world: how a new state is constructed out of previous ones given a

decision is expressed by a set of *belief rules*, denoted by B_i. Belief rules can conflict and agents can deal with such conflicts in different ways. The epistemic state therefore also contains an ordering on belief rules, denoted by $\geq_i^B$, to resolve such conflicts.

Definition 3 (Epistemic states). *Let a rule built from a set of literals be an ordered sequence of literals* $l_1, \ldots, l_r, l$ *written as* $l_1 \wedge \ldots \wedge l_r \to l$ *where* $r \geq 0$. *If* $r = 0$, *then we also write* $\top \to l$. *The* epistemic state *of agent* a_i, $1 \leq i \leq n$, *is* $\sigma_i = \langle B_i, \geq_i^B \rangle$, *where* B_i *is a set of rules;* $\geq_i^B$ *is a transitive and reflexive relation on the powerset of* B_i *containing at least the subset relation.*

Example 1. Let $s = \{p\}$ be the current state, $d_1 = \{a\}$, $B_1 = \{a \to q, a \wedge p \to \neg q\}$ and $\geq_1^B = \{a \wedge p \to \neg q\} > \{a \to q\}$: q is a consequence of action a, unless p is true: the second rule is an exception to the first one. The new state resulting from the decision d_1 in state s given the belief rules B_1 is $\{p, \neg q\}$: the applicable rules are $\{a \to q, a \wedge p \to \neg q\}$, but since they are conflicting only the rule $a \wedge p \to \neg q$ with higher priority in the ordering $\geq_1^B$ is applied.

The agent's motivational state contains two sets of rules for each agent. *Desire* (D_i) and *goal* (G_i) *rules* express the attitudes of the agent a_i towards a given state, depending on the context. When facing a conflict between their motivations, different agents prefer to fulfill different goals and desires. We express these agent characteristics by a priority relation on the rules which encode, as detailed in Broersen *et al.* [11], how the agent resolves its conflicts.

Definition 4 (Motivational states). *The* motivational state M_i *of agent* a_i $1 \leq i \leq n$ *is a tuple* $\langle D_i, G_i, \geq_i \rangle$, *where* D_i, G_i *are sets of rules,* $\geq_i$ *is a transitive and reflexive relation on the powerset of* $D_i \cup G_i$ *containing at least the subset relation.*

The decision process of an agent a_i tries to minimize (according to the ordering $\geq_i$ on goal and desire rules) the goal and desire rules in G_i and D_i which remain unsatisfied given a certain decision d_i.

Definition 5 (Unfulfilled motivational states). *Let* $U(R, s)$ *be the unfulfilled rules of state* s $U(R, s) = \{l_1 \wedge \ldots \wedge l_n \to l \in R \mid \{l_1, \ldots, l_n\} \subseteq s \text{ and } l \not\in s\}$ *The* unfulfilled mental state description *of agent* a_i *is* $U_i = \langle U_i^D = U(D_i, s), U_i^G = U(G_i, s) \rangle$.

Example 2. Given $\langle D_1 = \{\top \to z\}, G_1 = \{\top \to x, y \to w, z \to u\}, \geq_1 \rangle$ as the motivational state of agent a_1, the unfulfilled motivational state of agent a_1 in state $s = \{x, y\}$ is $U_1 = \langle U_1^D = \{\top \to z\}, U_1^G = \{y \to w\} \rangle$

In calculating which are the effects of a decision d_i given an initial state s, the agent uses the belief rules B_i and the ordering on them $\geq_i^B$ to resolve the possible conflicts. Moreover, agent a_i must predict the decisions of the agents acting after itself by recursively modelling ([13]) them using the information on their belief, goal and desire rules captured by their motivational states. The reader can find the details of the qualitative decision model in [5].

3.2 Norms

Prohibitions and permissions are defined in terms of goals and desires of the bearer of the norm and of the normative role, together with two auxiliary concepts. The first concept is *violation*. The normative role can decide whether something is considered a violation or not.

Definition 6 (Violation variables). *The violation variables of agent $p(a_j)$ are a subset of the decision variables of $p(a_j)$ written as $V_j = \{V_j^i(x) \mid x$ a literal built out of a propositional variable in $P \cup A_i \}$: x is a violation by agent $c(a_i)$.*

The second concept is *sanction*. Since it is not possible to assume that all agents are cooperative and respect the norms, sanctions provide motivations to fulfill the norms. A sanction is an action negatively affecting an agent, i.e., the agent desires the absence of the sanction.

Definition 7 (Conditional prohibition with sanction). *Agent $c(a_i)$ is prohibited by agent $p(a_j)$ to decide to do x (a literal built out of a variable in $P \cup A_i$) with sanction s (a propositional variable) under condition q (a proposition), $F_{(i,j)}(x, s \mid q)$, iff:*

1. *$q \rightarrow \neg x \in G_j$: if agent $p(a_j)$ believes that q it has as a goal that agent $c(a_i)$ adopts $\neg x$ as its decision.*
2. *$q \wedge x \rightarrow V_j^i(x) \in G_j$: if agent $p(a_j)$ believes that $q \wedge x$ then it has the goal $V_j^i(x)$: to recognize x as a violation done by agent $c(a_i)$.*
3. *$V_j^i(x) \rightarrow s \in G_j$: if agent $p(a_j)$ decides $V_j^i(x)$ then it has as a goal that it sanctions agent $c(a_i)$.*
4. *$\top \rightarrow \neg s \in D_i$: agent $c(a_i)$ has the desire not to be sanctioned.*

A permission to do x is an exception to a prohibition to do x if agent $p(a_j)$ has the goal that x does not count as a violation under some condition.

Definition 8 (Conditional permission). *Agent $c(a_i)$ is permitted by agent $p(a_j)$ to decide to do x (a literal built out of a propositional variable in $P \cup A_i$) under condition q (a proposition), $P_{(i,j)}(x \mid q)$, iff $q \wedge x \rightarrow \neg V_j^i(x) \in G_j$: if agent $p(a_j)$ believes $q \wedge x$ then it wants that x is not considered a violation done by agent $c(a_i)$.*

The permission overrides the prohibition if the goal that something does not count as a violation ($q \wedge x \rightarrow \neg V_j^i(x)$) has higher priority in the ordering on goal and desire rules $\geq_j$ with respect to the goal of a corresponding prohibition that x is considered as a violation ($q \wedge x \rightarrow V_j^i(x)$): $\geq_j \supseteq \{q \wedge x \rightarrow \neg V_j^i(x)\} > \{q \wedge x \rightarrow V_j^i(x)\}$. We do not consider here the problem of how normative role's characteristics can be generated; e.g., see [14] for a discussion of the problem of the legal sources of norms.

3.3 Resource, Authorization and Delegation

We introduce now the notion of resource, of control of a resource, of authorization and delegation of the institutional power to authorize access to a resource. An agent who manipulates a resource by means of some action is called a *resource consumer*:

Definition 9 (Resources). *Let RS be a set of resources. Let $RA_j = \{f_j(r) \mid r \in RS\}$ be a set of resource actions of agent $c(a_j)$ on $r \in RS$.*

The possibility to punish violations by means of some sanction s is among the pre-conditions for creating a prohibition; for this reason, it appears in the notion of controlling a resource, which is a precondition for issuing norms concerning access control. An agent who controls a resource is a *resource provider*.

Definition 10 (Control of resource). *Agent $p(a_j)$ controls a resource action f_i of agent $c(a_i)$ on resource $r \in RS$, $control_j(f_i(r))$, iff Agent $p(a_j)$ can negatively influence agent $c(a_i)$ when it executes $f_i(r)$ by means of some decision variable or parameter which it can control $s \in A_j \cup P$ such that $\top \rightarrow \neg s \in D_i$: agent $c(a_i)$ desires not to be sanctioned.*

As a particular case, $s = \neg p$ can be a literal built out of a parameter representing the failure of accessing a resource: e.g., reading a file has the desired effect of knowing the content of the file, and blocking the reading action results in the impossibility of knowing the information contained in the file. $c(a_i)$ believes that $p(a_j)$ with $m \in A_j$ prevents to achieve the effect p of $f_i(r)$ which $c(a_i)$ desires; $f_i(r) \rightarrow p \in B_i$, $\top \rightarrow p \in D_i$ and m has the effect $\neg p$: $m \rightarrow \neg p \in B_i$ and $\geq_i^B \supseteq \{m \rightarrow \neg p\} > \{f_i(r) \rightarrow p\}$.

Besides issuing norms, an agent which controls a resource can consider other agents authorized to access the resource it controls; authorizations, are a legal classification of reality for agents, and, thus, are represented by institutional facts:

Definition 11 (Authorizations). *Let the institutional facts I contain a set of so-called authorization variables: $H_j = \{u_j(f_i(r)) \mid a_i \in \mathcal{A}$ and $f_i(r) \in RA_i$ and $control_j(f_i(r))\}$ They are institutional facts representing that the resource provider $p(a_j)$ considers agent $c(a_i)$ authorized to access r with action f_i. An authorization has a meaning only if it appears among the conditions of a permission.*

Instead, declaring an agent authorized does not have the requirement to control a resource.

Definition 12 (Declarations). *Let the decision variables of agent $u(a_k)$ contain a set of so-called declaration variables $T_k = \{g_k(f_i(r)) \mid a_i \in \mathcal{A}$ and $f_i(r) \in RA_i\}$. Here $g_k(f_i(r))$ means that agent $u(a_k)$ declares agent $c(a_i)$ authorized to access r with action f_i.*

The point of declaring agents authorized is that a declaration generates an actual authorization if it counts as an authorization for the normative role controlling the resource. An example of this relation is the fact that a signature by the head of the department on a purchase order counts as the institutional commitment of the department to pay for that order: the head of the department has the institutional power to buy on behalf of the department.

Definition 13 (Counts as relation). *A decision variable $x \in A_k$ of agent $u(a_k)$, counts-as q, where q is a literal, for agent $p(a_j)$, $counts$-$as_j(x, q)$, only if $x \rightarrow q \in B_j$: agent $p(a_j)$ believes that x has q as a consequence.*

An agent who has been delegated the institutional power to authorize access is called an *authority*. It is not requested to control any resource.

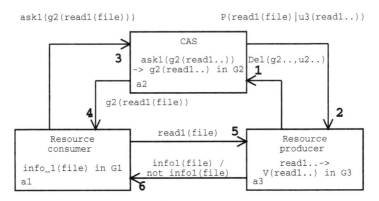

Fig. 1. Requesting access

Definition 14 (Delegation of authorization). *Agent $u(a_k)$ is delegated by agent $p(a_j)$ the institutional power to authorize agent $c(a_i)$ to do $f_i(r) \in RA_i$ by means of declaration $g_k(f_i(r)) \in T_k$ $(u_j(f_i(r)) \in H_j)$, $Del_{(k,j)}(g_k(f_i(r)), u_j(f_i(r)))$, iff we have counts-as$_j(g_k(f_i(r)), u_j(f_i(r)))$.*

3.4 Applicative Scenarios

In this section we sketch an applicative scenario in our model in the context of a virtual community. As shown in Figure 1 we have three agents: agent a_3, a provider $p(a_3)$ of resource f (a file), the resource consumer $c(a_1)$, and the CAS agent a_2: an authority $u(a_2)$. Agent $p(a_3)$ can block $c(a_1)$'s attempt of accessing with action r_1 ($r_1(f)$) the resource, since it is in control of the resource. When agent $p(a_3)$ joined the community (step 1) it maintained the prohibition to access f ($F_{(1,3)}(r_1(f), \neg info(f) \mid \top)$) but it agreed to share $r_1(f)$ the resource with the other members of the community by means of a permission to do $r_1(f)$. Unfortunately, $p(a_3)$ does not know which are the current members (in this case whether agent a_1 is a member) nor which is the access policy of the community concerning the resource f which $p(a_3)$ is sharing. Thus, agent $p(a_3)$ decides to consider what agent $u(a_2)$ says (or declares, in our terminology: $g_2(r_1(f)) \in T_2$) about $c(a_1)$'s access to f as an authorization $u_3(r_1(f))$ by itself: $u(a_2)$'s action $g_2(r_1(f))$ counts as $u_3(r_1(f)) \in H_3$ for $p(a_3)$. Then it permits agent $c(a_1)$ to access only if it is authorized: $P_{(1,3)}(r_1(f)|u_3(r_1(f)))$.

Agent $c(a_1)$ compares the different alternatives for achieving its goal of knowing the content of the file $info(f)$: requesting the resource by doing $r_1(f)$ alone (step 5) or first asking agent $u(a_2)$ for a declaration $(ask_1(g_2(r_1(f))))$(3) and then requesting to access the resource (5). It knows that a request for access is considered as a violation $V(r_1(f), c(a_1))$ by agent $p(a_3)$ and, thus, sanctioned by negating the information contained in the file $(\neg(info(f)))$. For this reason, it decides to ask agent $u(a_2)$ for an authorization to access agent $p(a_3)$'s resource. Agent $u(a_2)$ will provide $c(a_1)$ with the declaration since it is a goal of the $u(a_2)$ to cooperate with resource providers in enforcing the policy $(ask_1(g_2(r_1(f))) \rightarrow g_2(r_1(f)) \in G_2)$. Finally, agent $c(a_1)$ knows that the declaration $g_2(r_1(f))$ of agent $u(a_2)$ is considered as an authorization $u_3(r_1(f))$ by agent $p(a_3)$: $g_2(r_1(f)) \rightarrow u_3(r_1(f)) \in B_3$.

4   Related Work

This use of the terms right, authorization and permission as synonyms is frequent in policies for managing access control in distributed systems (e.g., [15]). In this paper we show why and how these concepts should be kept distinct in the context of virtual communities.

The necessity of a fine grained analysis of the concepts of permission and authorization in the security policy field is witnessed also by Sadighi Firozabadi and Sergot [16] who argue that a mere permission given by the resource provider to access a resource which it controls must be distinguished from the *entitlement* to access the resource: an agent is entitled and not merely permitted when the policies regulating the virtual community prohibit the resource provider not to permit the agent to access the resource.

Another distinction comes from deontic logic. Jones and Sergot [17] distinguish permissions from powers in the sense of having been delegated the *institutional power* to do something: "when we say that the Head of Department is authorized[1] to purchase equipment, we mean first and foremost that he has been granted by the institution the power to enter valid purchase agreements". Instead, "sometimes when we say that an agent is authorized to do such-and-such we mean no more than that he has been granted permission to do it".

Law studies argue that a further distinction must be drawn also in this last sense of the term authorization as mere permission. The [8]'s dictionary of law argues that adding or removing an authorization does not change the normative status of an agent while a new permission does; i.e., authorizations do not change the norms (prohibitions and permissions), an agent is subject to; rather, authorizations lift the legal obstacles and limitations, thus legitimating an action of the agent: they change the sphere of what is prohibited or permitted to the agent without adding or removing norms. This is possible since norms have a conditional character so that what is currently considered as a violation or not depends on which are the norms that have their conditions satisfied in the current situation. The fact that authorizations do not modify the existing norms, but change what is prohibited and permitted to an agent anyway, means that authorizations enable the conditions of some permissions. Hence, the institutional power to authorize can be delegated to other agents who do not directly control the resources, since creating an authorization does not require to change prohibitions and permissions.

Some scholars argue, instead, that the power to create permissions can be delegated. [18], for example, propose a framework where this power can be delegated as any other power to create institutional facts. We show in this paper that once prohibitions and permissions are not considered as primitive logical entities, the preconditions for their creation emerge. When we define them in terms of goals of the normative role, it emerges that controlling a resource is necessary for issuing a norm.

In Section 1, we highlighted that according to Cambridge Advanced Learner's Dictionary [7] *officiality* seems to be the first dimension distinguishing permissions from authorizations: the official character of authorizations depends on the fact that they are

---

[1] Note that Jones and Sergot use the term "authorization" in another sense with respect to this paper, i.e., as a synonym of "having the institutional power".

institutional facts, and this character distinguishes them from permissions; an authorization is an institutional fact which appears among the conditions of some permission issued by a normative role: if the normative role believes this fact, the permission is enabled so that what is permitted in the current situation is changed by the authorization. The creation of institutional facts is a commonplace feature of legal systems and norm-governed organizations. According to [17], "it is that particular agents are empowered to create certain types of states by mean of the performance of specific types of acts. Typically, the states created will have a normative character".

An authority has been *delegated the power to create an institutional fact* if the institution recognizes the authority's action as *counting as* something else (as in Searle [10]'s notion of construction of social reality). E.g., the fact that an authority declares an agent authorized counts as an authorization by a normative role. For Jones and Sergot [17], the counts as relation expresses the fact that a state of affairs or an action of an agent "is a sufficient condition to guarantee that the institution creates some (usually normative) state of affairs". They suggest this relation can be considered as "constraints of (operative in) [an] institution", and they express them as conditionals embedded in a modal operator.

5   Summary

We discuss policies for virtual communities based on peer-to-peer systems and the grid infrastructure with a community authorization service (CAS). Pearlman *et al.* [1] use the term 'right' both for the authorizations provided by the CAS and the permissions granted by the resource providers: "the user effectively gets the intersection of the set of rights granted to the community by the resource provider and the set of rights defined by the capability granted to the user by the community." We base our model on the distinction of the notions of permission and authorization, which leads to three roles for each agent: as a resource provider, a resource consumer and authority. The role played by the CAS in virtual communities is formalized in terms of what we called the authority role.

The task of *authorizing* requests performed by the authority CAS is not identical to the task performed by a resource provider when it *permits* access. Permissions and authorizations are distinguished, because authorizations can be delegated. They are related to prohibitions by the resource providers. The relation between the CAS and resource providers is that resource provides can be sanctioned. A resource provider can delegate to the CAS the power of authorizing resource consumers by declarations. The power to issue permissions cannot be delegated, because issuing permissions is restricted to control of the resource.

There are several issues for further research. First, delegation of authorization should be regulated,since not all authorities can be authorized themselves. Second, we can model hierarchies of policies to represent norms issued by local resource providers which can be constrained by obligations and permissions posed at the global level [19,20]. In [5] we discuss the counts as relation and constitutive rules in normative systems. Finally, in [21] we explore how to formalize our model using the standard BDI$_{CTL}$ logic [22] for agent verification.

References

1. Pearlman, L., Welch, V., Foster, I., Kesselman, C., Tuecke, S.: A community authorization service for group collaboration. In: Procs. of the IEEE 3rd International Workshop on Policies for Distributed Systems and Networks. (2002)
2. Moffett, J., Sloman, M.: Policy hierarchies for distributed systems management. IEEE Journal of Selected Areas in Communications 11(9) (1993) 1404–1414
3. Samarati, P., De Capitani di Vimercati, S.: Access control: Policies, models, and mechanisms. In Focardi, R., Gorrieri, R., eds.: Foundations of Security Analysis and Design LNCS 2171. Springer Verlag, Berlin (2001)
4. Sadighi Firozabadi, B., Sergot, M.: Contractual access control. In: Procs. of Workshop of Security Protocols, Cambridge (UK) (2002)
5. Boella, G., van der Torre, L.: Regulative and constitutive norms in normative multiagent systems. In: Procs. of KR'04. (2004) 255–265
6. Jin, X., Liu, J.: The dynamics of peer-to-peer tasks: an agent based perspective. In: Procs. of Agents and Peer-to-peer Computing. (2004) 84–95
7. Press, C.U.: Advanced Cambridge Learners' dictionary. Cambridge University Press (2002)
8. del Giudice, F.: Nuovo dizionario giuridico. Simone Editore (2001)
9. Makinson, D.: On the formal representation of rights relations. Journal of philosophical Logic 15 (1986) 403–425
10. Searle, J.: The Construction of Social Reality. The Free Press, New York (1995)
11. Broersen, J., Dastani, M., Hulstijn, J., van der Torre, L.: Goal generation in the BOID architecture. Cognitive Science Quarterly 2(3-4) (2002) 428–447
12. Lang, J., van der Torre, L., Weydert, E.: Utilitarian desires. Autonomous Agents and Multi-agent Systems (2002) 329–363
13. Gmytrasiewicz, P.J., Durfee, E.H.: Formalization of recursive modeling. In: Procs. of IC-MAS'95. (1995) 125–132
14. Boella, G., van der Torre, L.: Permissions and obligations in hierarchical normative systems. In: Procs. of ICAIL'03, ACM Press (2003) 109–118
15. Li, N., Grosof, B.N., Feigenbaum, J.: Delegation logic: A logic-based approach to distributed authorization. TISSEC 6(1) (2003) 128–171
16. Sadighi Firozabadi, B., Sergot, M.: Power and permission in security systems. In: Security Protocols. Springer Verlag (1999) 48–53
17. Jones, A., Sergot, M.: A formal characterisation of institutionalised power. Journal of IGPL 3 (1996) 427–443
18. Sadighi Firozabadi, B., Sergot, M., Bandmann, O.: Using authority certificates to create management structures. In: Procs. of Workshop of Security Protocols. Volume 2467., Berlin, Springer Verlag (2001) 134–145
19. Boella, G., van der Torre, L.: Local policies for the control of virtual communities. In: Procs. of IEEE/WIC WI'03, IEEE Press (2003) 161–167
20. Boella, G., van der Torre, L.: Local vs global policies and centralized vs decentralized control in virtual communities of agents. In: Procs. of WI'04. (2004)
21. Boella, G., van der Torre, L.: Game specification in normative multiagent system: the trias politica. In: Procs. of IAT'04. (2004)
22. Rao, A.S., Georgeff, M.P.: Decision procedures for BDI logics. Journal of Logic and Computation 8 (1998) 293–343

On Exploiting Agent Technology in the Design of Peer-to-Peer Applications

Steven Willmott, Josep M. Pujol, and Ulises Cortés

Universitat Politècnica de Catalunya,
Llenguatges i Sistemes Informàtics,
Campus Nord, Módul C5-C6, C/Jordi Girona 1-3, Barcelona (08034), Spain
{steve, jmpujol, ia}@lsi.upc.es

Abstract. Peer-to-peer (P2P) architectures exhibit attractive properties for a wide range of real world systems. As a result they are increasingly being applied in the design of applications ranging from high-
capacity file sharing and global scale distributed computing to business
team-ware. The objective of this paper is to outline a number of areas
in which Agent techniques for the management of *social problems* such
as decision making or fair trading amongst autonomous agents could be
used to help structure P2P actions. In particular we focus on approaches
from mechanism design, argumentation theory and norms / rules and
electronic institutions.

1   Introduction

Peer-to-peer (P2P) architectures exhibit attractive properties for a wide range
of real world systems. As a result they are increasingly being applied in the
design of applications ranging from high-capacity file sharing and global scale
distributed computing to business team-ware.

In addition their benefits however, P2P systems also fundamentally change
the networking paradigm used in an application often causing tensions with
other application goals such as security, predictability, performance guarantees,
billing and so forth. Some of these issues in particular arise due to the nature of
control, authority and *ownership* typical found in peer-to-peer systems:

- It is no longer possible to know exactly who is participating in the system.
- Participants in the system may change over time appearing and disappearing
  without trace.
- There are generally no centrally controlled 'arbiters' available to make au-
  thoritative decisions.
- Nodes may not only fail - they may be actively trying to subvert the system.
- Nodes may not only behave maliciously by themselves - subgroups of them
  may do so in a coordinated manner.

Each of these problems is not only technical but also *social* in nature [12] -
springing from the new found *autonomy* and *decision making power* of the peers
(actors) in the system. Whilst standard distributed systems engineering provides

G. Moro, S. Bergamaschi, and K. Aberer (Eds.): AP2PC 2004, LNAI 3601, pp. 98–107, 2005.

some of the answers, much work relevant work can also be found in the Agent and Multi-Agent Systems literature. The objective of this paper is to explore how some of these more 'social' P2P network issues could be addressed using various paradigms / approaches from the Agent research community.

While there are already many good examples of the application of Agent technology to P2P systems (such as many of the papers in previous editions of the Agents and Peer-to-Peer Computing workshop itself), with the exception of studies in the area of reputation and coordination [3] and others) the majority of this work to date has focused primarily on *algorithmic* or *infrastructural concerns*. In order to broaden this debate, in this paper we look at a number of other areas of Agent technology which could also bring significant benefits but have not been extensively applied to P2P systems as yet, these are mechanism design, argumentation theory and notions of norms / laws and electronic institutions. The paper is organised as follows:

- Section 2 briefly outlines some often positively and negatively perceived properties of P2P architectures.
- Section 3 characterises typical P2P systems in terms of different potential types of agent systems.
- Section 4 provides a number of example potential areas in which different types of Agent technology might benefit P2P application design.
- Section 5 concludes the paper.

The paper is discursive in nature and is intended to act as a discussion starter rather than an in-depth analysis of the issues involved.

2   The Good, The Bad and The Anti-social

While a certain amount of the interest in using P2P architectures in application development might be attributable to *hype* or *buzz*, they clearly also present key technical advantages beyond traditional client-server approaches for some applications. Some of the most visible of these advantages include:

- *Virtualised / transparent access to large-scale of distributed resources* - in particular computing resources. The SETI@Home search for extraterrestrial life program being one of the best known examples. [1]
- *Low configuration / low maintenance application deployment through Self organisation* - such as the easy to install and maintain teamware applications targeted by Groove Networks. [2]
- *High availability and fault tolerance* through replication, distribution or the extreme resilience of of power-law / scale-free topologies [3] such as the high-

---

[1] http://setiathome.ssl.berkeley.edu/

[2] http://www.groove.net/

[3] Power law topologies are highly resistant to random errors (failures) for example, although they can be more sensitive against directed errors (attacks). The high clustering coefficient of power-law networks also favours the redundancy of connections while improving communication: network diameter and average path length grow as the log function of the size (number of nodes).

capacity content caching achieved by services such as AKAMAI and Bit-Torrent. [4]

- *Anonymity* for users and providers - such as the information sharing services provided by the Freenet system which obfuscates the provider of information by sharing it between many hosts. [5]
- *Explosive deployment and growth* through peer download and installation - such as the extremely rapid user growth exhibited by new services such as Skype (a new-entrant global P2P voice-over-IP telephony service). [6]

Each of these advantages could provide a *decisive* business advantage in certain types of applications - allowing an enterprise to exploit a market and/or establish itself in a way that would be impossible with conventional client server approaches.

Inevitably however, adopting a P2P paradigm may also subject the subsequent application to a number of less desirable properties. Specific, well documented issues include:

- *Management challenges* - the deployed application can no-longer be directly managed as a global whole (raising issues on how to guarantee Quality of Service, perform maintenance / updates or even monitor its size).
- *Network fragmentation* - the network may become accidentally or deliberately subdivided parts which are not interconnected causing fragmentation of a service. Although service may degrade gracefully (showing robustness) by functioning in the remaining sub-parts, a service provider may loose control or contact with some parts of the network.
- *Identity issues* - the identity of users or systems connecting to the system may not be known (potentially raising issues of accountability for actions, fraud, trust and - in non-free services - of billing).
- *Security / Subversion* - malicious users or systems may be able to connect to system and subvert it by exploiting the lack of centralised authorities to monitor or control actions by its users.
- *User/Provider conflict* - the objectives of individual users / nodes may conflict with the global objectives of the network (e.g. users of file-sharing systems such as Gnutella benefit when finding files they would like, however there is no obvious motivation other than altruism / reputation for serving files).

While the first two issues might be considered standard networking or distributed systems issues the later issues are increasingly *social* in nature - since they depend on the nature, objectives and eventual actions of individuals using the application. Such problems arise in many classes of systems which are distributed (in ownership and/or space) and open - both of which hold in many P2P systems. These issues therefore raise important questions about the design

---

[4] http://bittorrent.com/
[5] http://freenet.sourceforge.net
[6] http://www.skype.com/

of P2P application that are generally not be present in their client-server approach. In some cases the issues may also make the choice of a P2P approach inappropriate.

The remainder of this paper is dedicated to looking at how we might use Agent technologies to help exploit some of the benefits of P2P systems whilst mitigating or managing the downsides.

3   Peer-to-Peer Systems as Agent Systems

Before addressing which areas of Agent technology might be applicable to P2P systems it seems worth spending some time examining the properties of P2P systems in terms of typical Agent characterisations.

Informally, nodes in a P2P network can be characterised as fulfilling many current definitions of Agenthood (see [14] for an overview) to a greater or lesser degree. Arguably more important than the properties of each peer as an agent however, are considerations of what type of *Multi Agent System* the application corresponds to.

In the general case one would expect that:

- Peers are *entirely autonomous:* each individual peer could act in any way - conceivably any code provided could have been entirely re-engineered by its owner/user.
- However, peers are *bound to a specific limited set of actions, protocols or messages* specified in the protocols defined for the application - that is they are limited to an agreed set of social conventions which may be broad or narrow.
- *Rational behaviour cannot be guaranteed:* an owner/user may have any number of *external* motivations for particular actions - some of which may not correspond to *rational actions* in the system itself. On the other hand rational action can be made more likely if:
  - Significant participation costs are involved (or in particular if irrational actions have direct costs).
  - Commitments made during participation can be enforced.
- *Cooperative behaviour cannot be assumed:* the motivations for actions of an individual user/peers are *unlikely to be primarily for the social good* - but primarily for that user/peers' own good. Obvious examples include the phenomenon that systems such as Gnutella are dominated by so called *free riders* [1] - the assumption of benevolence cannot be made.
- *Out-of-band coalition formation and/or collusion is possible and likely:* in other words users/systems are likely to use additional communication channels invisible to application to coordinate their actions in groups when and where this is of benefit.
- *False name / identity participation is possible:* in other words users may create multiple identities (new P2P nodes) to participate in the system if this could lead to financial/other benefits - such as influencing market prices or manipulating trust/other social properties.

These properties unfortunately correspond to problems recognised as *some of the most complex and difficult to deal with* in the Multi-Agent research literature. An analysis of the Multi-Agent Systems literature would show that only a relatively small percentage of known results are directly valid for these conditions.

Hence for applications which fit the above profile, engineering coherent behaviour amongst users of the P2P application is likely to be very challenging. On the more positive side in some cases it might be possible to make some additional assumptions on average over the whole population:

- *Average rationality:* that *on average* nodes/users act rationality, although an individual may not (this could be justified for example in large market scenarios with many users).
- *Verifiable identity:* that false name / fake identity problems can either be excluded or at least made very rare (this could be justified in applications which directly tie application participation to some other verified identity mechanism such as digital certificates and/or corporate employee registrations).

In other applications it may be possible to reduce the impact of other characteristics. However in each application case it is important to capture the assumptions which do and do not hold since they may fundamentally affect the correct functioning of the system.

4   Where Can Agent Technologies Help?

In order to illustrate how Agent technologies might help in P2P application design this section presents application examples coupled with a description of how particular techniques from the Agent literature might be used.

The descriptions are intended to be examples only; however we hope that they will be useful as food for thought for other potential domains / technology application. The examples are:

1. Mechanism Design *applied to* P2P trading systems.
2. Argumentation / Negotiation schemas *applied to* P2P-Social Choice problems.
3. Electronic Institutions, Norms, Rules, Policy Languages *applied to* context management in P2P ubiquitous computing problems.

Other diverse examples could include reputation and trust applied to social networking systems or Agent Communication Language semantics applied to interoperability amongst peers.

4.1   Peer-to-Peer Trading Systems and Mechanisms Design

P2P applications show great potential for trading or bargaining systems in which users are able to purchase / exchange goods, services, information or other items.

Currently most well known Internet market applications such as EBAY for consumers and a myriad of Business-to-Business trading systems are strongly centralised in nature; however there are strong reasons why P2P approaches might be attractive: dis-intermediation of middle-players (potentially cutting out fees), privacy (since no one entity knows about all transactions), robustness and speed (reducing dependency on a single site) and others.

Unfortunately however, unlike in file-sharing systems where goods/services are provided at near zero loss to the provider, trading systems require real economic exchange. The subsequent potential for financial loss raises the obvious question: *how do we ensure that the rules in the application cannot be violated or abused by one or more parties to defraud other parties?* Standard responses to this such as requiring users to register their identities or the appointment of arbiters to regulate disputes quickly become unmanageable, can themselves be defrauded (e.g. by identity theft) or begin to recreate centralised elements that the application aimed to removed.

Although still in its infancy for strongly distributed systems the research area of mechanism design [8] which brings together theories from economics and multi-agent systems is extremely relevant here. Work by researchers such as Toumas Sandholm and David Parkes traces the boundaries of:

- Which properties market systems conforming to a particular combination of rules and protocols exhibit.
- Which strategies dominate under which conditions (and whether they are desirable strategies such as telling the truth about valuations for goods or not).
- The impact of potential market elements such as side payments (additional financial exchange occurring between agents in a market which is not openly carried out using the market itself).

4.2 Argumentation Based Negotiation Applied to Social Choice Problems

A specific example of mechanism design which does not necessarily involve financial exchange is in the solution of social-choice problems. In these applications, peers are required to provide opinions on a question such as the passing of a law, the truth of a statement, the election of a particular agent to a social role or something similar. This social choice generally needs to be made in such as a way as to respect various definitions of fairness, speed (termination) and proof against manipulation.

Mechanism design provides tools which analyse the properties of different types of voting systems (see [4,5] for example). However, in cases where the full mechanism cannot be analysed, the decisions cannot be made by simply *weight of opinion* or the choices are not simple to enumerate (e.g. in deciding an arbitrary set of facts that is true) other techniques may also help. The challenge in such applications becomes: *how can I structure interactions between peers in order to reach a fair compromise?*

An emerging area which could be applied here is that of Argumentation theory [15,10,17] which provide methods and formalisms to structure dialogues and conversation rules amongst agents in order to ensure properties such as:

- Valid deduction of facts, commitments or other statements from assertions made by agents.
- Termination of the interaction.
- Allowing all agents to have their say - but preventing a vocal few drowning out discussion.

Papers range of different ways of formalising conversation rules [18] to applications in choice problems such as systems for guiding democratic debate.

4.3 Ubiquitous Computing: Institutions, Organizations, Policies, Rules and Norms

A further often cited application of P2P networked is the growing trend of viewing everyday objects (household appliances, mobile devices, clothes, vehicles or almost anything) as augmented communicating devices - each with its own identity and networked functionality. Generically known as *Ubiquitous Computing*, this vision relies on objects being fitted with small computational devices, network capability and behaviour patterns which provide their owners with additional benefits such as information, control, remote activation and so forth [11].

The potential explosion in such devices and the complexity of their interactions means that client-server architectures are generally expected to be overwhelmed and unable to provide reasonable management of such systems. In particular challenges arise such as: *given so many devices - how can we ensure they do not clash with one another? How can we manage behavioural changes between (for example) home, office and street? How can ensure particular behaviour is enabled/disabled in situations when it may be dangerous?*

A concrete example might be visiting a neighbour's house bringing a wireless enabled device - and having it automatically interact with the house systems / objects [7] - changing its status to avoid clashes and being treated with care by the local network as a potential security threat (e.g. unbeknown to the owner it may have been infected with a computer virus). The resulting interactions are highly non-trivial and in particular dependent on the context of the interaction (home, office, street), the current world state (time, weather, malicious 3rd party network activity), the device owners' relationship and many other factors.

Whilst some relationships, rules and heuristics could be hard coded (such as a process for deciding whether a new device is permitted to access the network), as the number of rules and interactions grows the management problem looks set to explode. Several areas of Multi-Agent Systems research can provide useful tools in this respect:

- Techniques for modelling Norms, Rules, Laws an Electronic Institutions [9,16].

---

[7] See [6] for an office example.

- Coordination techniques based on declarative rules such as Shoham's Social Laws [20].
- Work on policy languages and models ([7] for example).

Each of these techniques describes collections of rules (from abstract to concrete), properties and other constraints on populations of agents: creating a tangible social context which systems are governed by. Most importantly the approaches are primarily declarative - allowing easy management of the rules and the derivation of guaranteed properties for certain combinations of rules. In a Ubiquitous Computing P2P system the approaches could be used to assign contextual roles to peers, tracking their obligations and rights over time as they change context (through motion or changes in the owner's attitude for example). Hence all devices are peers at one level (objects all have an identity and the ability to act) but are dynamically structured into organisational structures according to need.

This last scenario can also be seen as a superset of the previous two in that market mechanism rules and/or argumentation schemas are often seen as defining an institutional context (or applied in the context of a particular institution).

5   Conclusions

The abstract for the panel discussion held at the P2P-Agents workshop in New York stated a number of challenges which the P2P systems (in particular for business) faced including *security, trust and reputation, representing business protocols, checking compliance, bootstrapping systems, and performance*. On closer analysis, P2P systems can arguably be recognised as having similar characteristics as some of the hostile environments described in the Multi-Agent Systems literature: peers are entirely autonomous, individual autonomy cannot be assumed, out-of band collusion between peers is possible and so forth.

However even under these conditions work in areas such as Mechanism Design and Norms/Institutions or structured Negotiation can potentially provide tools for engineering predictable applications by analysing:

- The motivations of actors.
- The implicit and explicit social rules and properties of the systems.
- Which types of behaviour can monitored/guaranteed and which not.
- The relationship between in-band actions/interactions (those which form part of the application) and those which might take place out-of-band (externalities or group collusion for example).

Work on reputation in Agent based P2P systems already follows these lines ([3,13] and others) but we hope that the contents of the paper helps illustrate that there may also be other elements of Multi-Agent system's theory which can address these more *social* P2P application issues.

Acknowledgements

This work was partly supported by the European Projects ASPIC [8], Provenance (IST FP6-511085) and @lis technology net [9]. The work also owes much to discussions with colleagues in the Agentcities and openNet Initiatives.

Notwithstanding this the opinions expressed in the paper are those of the authors and do not necessarily reflect those of other project participants.

Particular thanks go to the other panellists in the session (Sonia Bergamaschi, Sandip Sen, Hector Garcia Molina) and the panel chair Munindar Singh.

References

1. Adar, E. and Huberman, B. A.: Free Riding on Gnutella Technical Report Xerox Laboratories, 2000, Internet Ecologies Area. http://www.hpl.hp.com/research/idl/papers/gnutella/gnutella.pdf
2. Atkinson, K., Bench-Capon, T. and McBurney, P.: PARMENIDES: Facilitating democratic debate. Third e-Government Conference (EGOV04), DEXA 2004, Zaragoza, Spain, September 2004. Published in: R. Traunmuller (Editor): Lecture Notes in Computer Science. Berlin, Germany: Springer.
3. Biswas, A., Sen, S. and Debnath, S.: Limiting Deception in a Group of Social Agents Applied Artificial Intelligence, 14:785–797, 2000.
4. Conitzer, V. and Sandholm, T.: Complexity of Manipulating Elections with Few Candidates. Applied Artificial Intelligence, 14:785–797, 2000. In Proceedings of the National Conference on Artificial Intelligence (AAAI), 2003.
5. Conitzer, V., Lang, J, and Sandholm, T.: How Many Candidates Are Needed to Make Elections Hard to Manipulate? Applied Artificial Intelligence, 14:785–797, 2000. In Proceedings of the Conference on Theoretical Aspects of Rationality and Knowledge (TARK), Bloomington, Indiana, 2003.
6. Chen, H., Finin, T., Joshi, A., Perich, F., Chakraborty, D. and Kagal, L.: Intelligent Agents Meet the Semantic Web in Smart Spaces IEEE Internet Computing, to appear, 2004.
7. Damianou, N., Dulay, N., Lupu, E. and Sloman, M.: The Ponder Policy Specification Language Lecture Notes in Computer Science, 1995.
8. Dash, R., Jennings, N., and Parkes, D.: Mechanism Design: A Call to Arms Computational, IEEE Intelligent Systems, November 2003, pages 40-47 (Special Issue on Agents and Markets).
9. Dignum, F.: Autonomous Agents with Norms Artificial Intelligence and Law 7:69-79, 1999.
10. Dung, P.: On the acceptability of Arguments and its fundamental role in the Nonmonotonic Reasoning, Logic Programming and N-Person Games Artificial Intelligence 77, 2, pp 321-358, 1995.
11. Fitzmaurice, G.W., Ishii, H. and Buxton, W.: Bricks: Laying the Foundations for Graspable User Interfaces CHI, pp 442-449, 1995.
12. Jennings, N. R.: Controlling Cooperative Problem Solving in Industrial Multi-Agent Systems using Joint Intentions Artificial Intelligence 75, pp 195-240, 1995.

---

[8] http://www.argumentation.org/
[9] http://www.alis-technet.org

13. Jurca, R. and Faltings, B.: Towards Incentive-Compatible Reputation Management. Trust, Reputation and Security: Theories and Practice, Lecture Notes in AI 2631, 2003, pp. 138-147.
14. Luck, M., McBurney, P. and Preist, C.: Agent Technology: Enabling Next Generation Computing January 2003, ISBN 0854 327886. http://www.agentlink.org/roadmap/index.html
15. Kraus, S., Nirkhe M. and Sycara, K.P.: Reaching agreements through argumentation: a logical model (Preliminary report) Proceedings of the 12th International Workshop on Distributed Artificial Intelligence, Hidden Valley, Pennsylvania, pp 233–247, 1993.
16. Panchoco, O and Carmo, O.: A role based model for normative specification of organised collective agency and agent interaction JAAMAS 6(2):145-183, 2003.
17. Parsons, S., Sierra, C. and Jennings, N.: Agents that Reason and Negotiate by Arguing Journal of Logic and Computation, 8, 3, pp 261–292, 1998.
18. Parsons, S., McBurney, P. and Wooldridge, M.: The mechanics of some formal interagent dialogues F. Dignum (Editor): Advances in Agent Communication. Lecture Notes in Artificial Intelligence 2922. pp 329–348. Berlin, Springer. g
19. Sandholm, T.: Automated mechanism design: A New Application Area for Search Algorithms In Proceedings of the International Conference on Principles and Practice of Constraint Programming (CP), 2003.
20. Shoham, Y. and Tenneholz, M.: On the synthesis of useful Social Laws for Agent Societies. In proceedings AAAI-92, pp 276-281, 1992

Peer-to-Peer Semantic Integration of XML and RDF Data Sources*

Isabel F. Cruz, Huiyong Xiao, and Feihong Hsu

Department of Computer Science,
University of Illinois at Chicago, USA
{ifc, hxiao, fhsu}@cs.uic.edu

Abstract. Peer-to-Peer (P2P) data management systems combine traditional schema-based integration techniques with the P2P infrastructure. In this paper, we propose a P2P data management framework named PEPSINT that semantically integrates heterogeneous XML and RDF data sources, using a hybrid architecture and a global-as-view approach. Our focus is on the query processing techniques over heterogeneous data. Queries in PEPSINT are expressed in XQuery and in RDQL. We consider two types of queries, depending on whether the query is first posed on the super peer or on one of the peers.

1 Introduction

The Semantic Web has been proposed to add semantics to web content and to enable interoperability among heterogeneous data sources. Both Extensible Markup Language (XML) and Resource Description Framework (RDF) can be used to represent information on the Web. However, there exists a wide gap between the two languages, since RDF data has *domain structure* (the concepts and the relationships between concepts) while XML data has *document structure* (the hierarchy of elements) [11].

An example is shown in Figure 1, in which the RDF schema R explicitly specifies two concepts, `Book` and `Publisher`, as well as the `publishedBy` relationship. Figure 1 also shows two XML schemas S_1 and S_2. Each of these XML schemas contains two concepts: `book` and `author` (equivalently denoted by `article` and `writer` in S_2). Conceptually, these two XML schemas are quite similar. Structurally speaking, however, they are very different: S_1 (book-centric schema) has the `author` element nested under the `book` element, whereas S_2 (author-centric schema) has the `article` element nested under the `writer` element.

Furthermore, the wide diversity of possible XML schemas for a single conceptual model also results in wide diversity for the XML queries. For instance, a user who wants to "List all the publications" from two data sources corresponding to S_1 and S_2 may write the XML path expressions, respectively, as `/books/book/@booktitle` and `/writers/writer/article/@title`. We notice

---

* This research was supported in part by the National Science Foundation under Awards EIA-0091489 and ITR IIS-0326284.

G. Moro, S. Bergamaschi, and K. Aberer (Eds.): AP2PC 2004, LNAI 3601, pp. 108–119, 2005.
© Springer-Verlag Berlin Heidelberg 2005

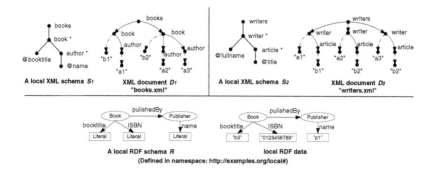

Fig. 1. An example of heterogeneous XML and RDF data sources

that although the two XML path expressions refer to semantically equivalent concepts, they follow two distinct XML paths. In contrast, schemas defined on the conceptual level (known as *conceptual schemas* or ontologies) are *flat* in document structure, and therefore the user can formulate a query without considering the structure of the source (we refer to such queries as *conceptual queries*). RDF Schema (RDFS), DAML+OIL, and OWL are examples of languages used to create conceptual schemas.

There are currently several attempts to use *conceptual schemas* [1,2,8,9] and *conceptual queries* [6,7] to overcome the problem of structural heterogeneities among XML sources. In this paper, we propose a framework called PEPSINT (PEer-to-Peer Semantic INTegration framework) to semantically integrate heterogeneous XML and RDF data sources in a P2P environment. We discuss the architecture of PEPSINT, and present a solution for semantic integration and query processing in the P2P heterogeneous environment. In brief, we make the following contributions in this paper:

- We propose a P2P schema-based data management framework, PEPSINT, built on a *hybrid* P2P architecture, in which the global RDF ontology (constructed using the global-as-view approach [13]) in the *super peer* behaves not only as a central control point over the *peers* but also as a mediator for query translation from peer to peer.
- For the purpose of semantic integration, we propose an approach that preserves the domain structure of RDF and the document structure of XML. Specifically, the semantic integration of XML and RDF data sources is implemented at the schema level (through the schema matching process) and at the instance level (through the query answering process).
- We also provide a set of query rewriting algorithms that can propagate a user's query across the heterogeneous XML or RDF data sources in PEPS-INT. In our framework, mappings connect the peer to the super peer, thus making query processing within the network transparent to a user in any peer.

The paper is organized as follows. Section 2 gives a review of related work. In Section 3 we describe the architecture of PEPS-INT and its main components.

Section 4 discusses schema-based integration of RDF sources and (structurally dissimilar) XML sources. Query processing in PEPSINT is covered in Section 5. Finally, we draw conclusions and discuss future work in Section 6.

2   Related Work

The research community has, to date, produced several P2P data management systems that aim to enable interoperability among distributed heterogeneous data sources.

The **Edutella** project [15] provides an RDF-based metadata infrastructure for P2P networks based on the JXTA framework [10]. In Edutella, connections between peers are encoded into a network topology known as the *Edutella super-peer topology*, which is similar to the hybrid architecture used in PEPSINT. A Datalog-based query exchange language called RDF-QEL is proposed to serve as a common query interchange format. Thus a wrapper translates local query languages such as SQL and XPath into RDF-QEL. Edutella does not support XML sources directly, though the RDF data sources may be serialized in XML format.

PeerDB [16] is an agent-based P2P data management system where each peer holds a relational database. The metadata for relations that are sharable with other peers is specified in a local *export dictionary*. Unlike PEPSINT, there are no established mappings between peers. Thus, query reformulation between peers in PeerDB is assisted by agents through a *relation-matching strategy*; this is a process of matching the metadata between relations in different peers. XML and RDF data are not considered in the current implementation of PeerDB.

SEWASIE [4] is another agent-based P2P system that aims to integrate Information Nodes (SINodes), where each node acts as an autonomous mediator-based system. It contains two types of agents: *query agents* that are responsible for query processing and answering; and *brokering agents* (peers) that handle the mappings between nodes. SEWASIE does not currently support RDF data sources.

Hyperion [3] proposes an architecture for a P2P data management system for relational databases (one stored at each peer). Similarly to PEPSINT, *mapping tables* and *mapping expressions* (mapping tables that allow variables) are used to store connections between local schemas in peers. Unlike PEPSINT, only relational data sources and relational queries are supported by Hyperion.

The **Piazza** system [11] is a P2P data management system that, like PEPSINT, supports interoperation of both XML and RDF data sources. Furthermore, both systems preserve document structure of XML sources during interoperation of these sources. The differences from PEPSINT are: (1) Piazza is based on the pure P2P architecture in which peers are connected directly, whereas PEPSINT is built on top of a hybrid architecture with a super peer containing the global ontology. This is a tradeoff between efficiency and autonomy [4]. (2) Piazza uses a (declarative) XQuery-based mapping language for mediating between nodes, whereas PEPSINT utilizes mapping tables to store schema correspondences, which we believe results in easier construction and maintenance of mappings. (3) The Piazza system achieves its interoperability in a low-level

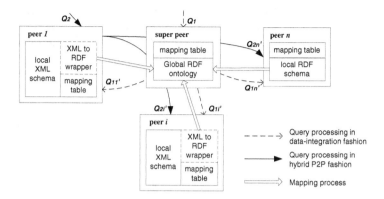

Fig. 2. The PEPSINT architecture

(syntactic) way, i.e., through the interoperability of XML and the XML serialization of RDF. For this reason, the user has to write an RDF query in terms of an XQuery. The query rewriting in Piazza is based on pattern matching between an XQuery expression and the mappings. In contrast, PEPSINT supports RDF queries at the conceptual level (RDQL), as well as XQuery. Query translation is realized by a collection of query rewriting algorithms.

3   The PEPSINT Architecture

There are two types of P2P architectures [14]: the *pure P2P architecture*, in which no central point of control exists and peers are autonomous but can communicate directly with each other; and the *hybrid P2P architecture* that contains at least one central point of control. The global control point(s) maintain either network control or the references to the remaining peers. Based on the hybrid P2P architecture, PEPSINT contains two types of peers: the *super peer*, containing the global RDF ontology, and the *peers*, containing local schemas and local data sources. Each peer represents an autonomous information system and connects with the super peer by establishing P2P mappings. As shown in Figure 2, the PEPSINT architecture has four main components.

XML to RDF Wrapper. Since XML is characterized by having a hierarchical document structure while RDF has a flat document structure, it is hard for the user to directly map a local XML schema to the global RDF ontology. To solve this problem, an *XML to RDF wrapper* is used to transform the XML schema into a local RDF schema, which is then mapped to the global ontology. This is a process that conceptualizes the XML elements into RDF concepts while keeping their nesting information (by using a specialized RDF property).

Local XML and RDF Schemas. The local XML and RDF schemas residing in peers contain both data and metadata. For the purpose of semantic integration, we represent a local RDF schema as a labeled digraph (from now on referred to as *RDF schema graph*). The domain structure is explicitly represented by labeled vertices (concepts) and labeled arcs (relationships between concepts). Likewise, a local XML schema is represented as a labeled tree (from now on referred to as *XML schema tree*) that specifies nesting relationships between labeled vertices (elements).

Global RDF Ontology. The global RDF ontology in the super peer is a virtual mediated schema integrated from distributed local RDF schemas (using the global-as-view approach [13]). In PEPSINT, the global ontology has two roles: (1) It provides the user with a uniform and complete view of data sources in the distributed peers; and (2) it serves as a mediator for query translation from one peer to other peers. The global RDF ontology is a fairly simple ontology—it does not contain high-level axioms, such as those available to DAML+OIL or OWL.

Mapping Table. A mapping table stores mappings between local schemas and the global ontology. Each mapping represents correspondences between concepts of different local schemas and is stored (as an entry) in the mapping table. It is easy to maintain the mappings by adding, deleting, or updating the entries. This feature of the mapping table makes it well fit the dynamic nature of P2P environments, in which data sources may be added or removed frequently. We use *XML path expressions* to represent the elements contained in an XML schema, and *RDF path expressions* to represent the concepts and relationships in an RDF schema.

The operation of PEPSINT can be divided into two phases: *mapping (or design) phase* and *query (or runtime) phase*, as respectively indicated by the hollow arrowed lines and the solid and dashed arrowed lines in Figure 2. To realize semantic integration of XML and RDF data sources, domain structure and document structure must be preserved in both phases.

1. Mapping Phase. Whenever a new peer joins the PEPSINT network, the peer gets registered and indexed in the super peer by establishing mappings from its local schema to the global ontology. The mappings are established through a process of *schema matching* [1] and stored in the mapping table of the peer. During the process of schema matching, the global ontology is extended by integration of the local schemas. As previously mentioned, the domain structure and document structure of local schemas are encoded in the mappings.

2. Query Phase. PEPSINT provides two query processing modes. (1) In the *data-integration mode*, the user poses a query (*source query*) on the global ontology in the super peer, which is then reformulated into multiple subqueries (*target queries*) over the XML and RDF sources in the peers (one subquery for each source). By executing the target queries and integrating their results, the system returns an answer to the user at the site of the super peer. (2) In the

---

[1] Schema matching is a basic problem in many database application domains, and currently it must be performed manually. A taxonomy covering most of the existing approaches to schema matching has been devised [17].

hybrid P2P mode, the user can pose a source query on the local XML or RDF source in some peer. Locally, the query will be executed on the local source to get a *local answer*. Meanwhile, the source query is reformulated into a target query over every other peer through transitive mappings (compositions of mappings from the original peer to the super peer and mappings from the super peer to the other target peers). By executing the target query, each peer returns an answer to the original peer, called the *remote answer*. The local and remote answers are integrated and returned to the user at the site of the originating peer.

Query translation is achieved by using the mappings in conjunction with a collection of query rewriting algorithms. We discuss the mapping and query phases in greater detail in Section 4 and Section 5, respectively. Running examples based on the schemas in Figure 1 will be used for illustration.

4   Mapping Process

In PEPSINT, the data sources residing at the peers may be either XML data modeled by an XML schema language (e.g., XML Schema) or else RDF data whose classes and properties are described using RDF Schema (RDFS). As previously mentioned, mappings between local schemas and the global ontology are established by the schema matching process during the registration of a peer to the super peer. The key operation in this process is the preservation of the domain structure of RDF sources and the document structure of the XML sources.

4.1   Mapping a Local RDF Schema to the Global RDF Ontology

Schema matching takes the global RDF ontology G (in the super peer) and a local RDF schema R (in the peer) as the inputs and returns a set of mappings M between the elements of G and the elements of R as the output. Meanwhile, the global ontology is updated by merging or adding metadata from the local RDF schema.

Elements in an RDF schema include *concepts* and *roles* (also known as *classes* and *properties* in RDFS terminology). When matching the local RDF schema with the global RDF ontology, for each element p_L in the local RDF schema, if there already exists in the global ontology a semantically equivalent element p_G, the two elements will be merged and a correspondence such as (p_L, p_G) will be generated. Otherwise, the element p_L will be copied into the global ontology

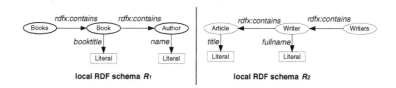

Fig. 3. RDF schemas transformed from the local XML schemas in Figure 1

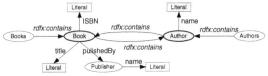

Global RDF ontology *G* (defined in namespace: http://examples.org/global#)

RDF path expressions in G	RDF path expressions in R	XML path expressions in S_1	XML path expressions in S_2
Books	–	/books	–
Book	Book	/books/book	/writers/writer/article
Book.title	Book.booktitle	/books/book/@booktitle	/writers/writer/article/@title
Book.ISBN	Book.ISBN	–	–
Book.publishedBy	Book.publishedBy	–	–
Publisher	Publisher	–	–
Publisher.name	Publisher.name	–	–
Authors	–	–	/writers
Author	–	/books/book/author	/writers/writer
Author.name	–	/books/book/author/@name	/writers/writer/@fullname

Fig. 4. The global RDF ontology and its mapping table

as p_G, and a correspondence (p_L, p_G) will be generated as well. In [9], we have defined a group of operations on the ontology to implement schema matching between two RDF schemas, e.g., *merging of classes*, *merging of properties*, *merging of relationships between classes*, and *copying a class and/or its properties*.

4.2   Mapping a Local XML Schema to the Global RDF Ontology

By transforming the participating local XML schema into a local RDF schema, we can convert the problem of matching an XML schema with the global ontology into the problem of matching an RDF schema with the global ontology, which is discussed in Section 4.1.

The schema transformation is carried out by the XML to RDF wrapper. The XML to RDF wrapper converts XML attributes and simple elements to RDF properties; it converts XML complex elements to RDF classes. The wrapper also encodes the element-attribute relationship and the element-subelement relationship in XML schema respectively as the *class-to-literal* relationship and the *class-to-class* relationship in the resulting RDF schema.

We choose to define a new, specialized RDF property *rdfx:contains* (the prefix *rdfx* stands for the new name space "http://pepsint.org/rdfx#") to explicitly denote nesting relationships. In particular, given that two XML elements e_i (parent element) and e_j (child element) are respectively converted into two RDF classes, c_i and c_j, the property *rdfx:contains* of c_i is then generated to connect c_i to c_j. Figure 3 shows the resulting local RDF schemas R_1 and R_2 that are respectively converted from the two XML schemas S_1 and S_2 shown in Figure 1. Finally, the global ontology G integrated from S_1, S_2 and R (in Figure 1) and its mapping table are shown in Figure 4. The grayed concepts or roles are the ones merged from local sources. We notice that both the *rdfx:contains* property in G

and the mappings in the mapping table encode the document structure of XML sources, so that either of them can be exploited for tracking XML document structure in future query translations.

5   Query Processing

5.1   Assumptions

For the simplicity of discussion, we make the following assumptions.

1. We assume the mappings from a local schema to the global ontology are total, one-to-one mappings. On the other hand, the mappings from the global ontology to the whole set of local schemas are total but not one-to-one mappings, since a concept in the global ontology might be merged from multiple concepts of different local schemas (as a result of schema matching). The mappings from the global ontology to a single local schema are one-to-one but they may be partial mappings, which means a query run at a local source may result in an incomplete answer.

2. We also assume that XML queries conform to a subset of XQuery [5], which we call PXQuery (*Partial XQuery*) in this paper. PXQuery consists of a non-nested FLWR expression that includes four clauses: `for`, `let`, `where`, and `return`; the `where` clause may only contain comparison operators. Other limitations of PXQuery include: (1) Only a single XML document is involved in the query; (2) No new XML fragments are introduced in the query; (3) The path expressions contained in the clauses only use child axes; (4) No type declarations, functions, `order` clauses, and predicate filters are used.

3. To represent RDF queries, we use RDQL, which uses an SQL-like syntax [12]. RDQL consists of the following clauses: `SELECT`, `FROM`, `WHERE`, `AND`, and `USING`. We assume only comparison operators are used in the AND clause of the RDQL query. The `FROM` and `USING` clauses are not the focus of our attention since they are not involved in query translation.

For the sake of convenience, we associate a PXQuery query Q with (V_{Q^R}, V_{Q^W}, C_Q), where V_{Q^R} and V_{Q^W} are the two sets that respectively contain all XML path expressions in the `return` clause and in the `where` clause, and C_Q contains the constraints whose items are in the form of vRc, where $v \in V_{Q^W}$, c stands for a constant, and R is a comparison operator (e.g., $=$, $<$, $>$, $\leq$, $\geq$, and $\neq$). Likewise, we also associate an RDQL query Q with a triple (P_{Q^S}, P_{Q^W}, C_Q), where P_{Q^S} and P_{Q^W} respectively contain all RDF path expressions in the `SELECT` clause and in the `WHERE` clause, and C_Q contains the constraints whose items are in the form of pRc, where $p \in P_{Q^W}$, c stands for a constant, and R is a comparison operator.

5.2   Query Answering in Data Integration Mode

Query answering in data integration mode includes the following steps. We use a running example for illustration.

1. **Analyzing the source RDQL query** to convert it from a string to a triple $Q_{in} : (P_{Q_{in}^S}, P_{Q_{in}^W}, C_{Q_{in}})$. In order to get the RDF path expressions in $P_{Q_{in}^S}$

and $P_{Q_{in}^W}$, we have to match the triple patterns (specified in the WHERE clause) with the RDF graph corresponding to the local RDF schema. $C_{Q_{in}}$ contains all the constraints specified in both the triple patterns of the WHERE clause and the AND clause. Because of space limitations, we ignore the detailed process of pattern matching in this paper.

Example 1. To "find the publications written by a1", the user poses a query over the global ontology as shown below on the left hand side (the prefix go stands for the name space "http://examples.org/global#", where the global ontology is defined). The resulting Q_{in} elements are listed on the right hand side.

```
SELECT ?title                              P_{Q_{in}^S}={Book.title}
WHERE (?book, <go:title>, ?title),         P_{Q_{in}^W}={Book, Book.title, Author,
      (?book, <rdfx:contains>, ?author),           Author.name}
      (?author, <go:name>, ?name)          C_{Q_{in}}={(Author.name, eq, "a1")}
AND (?name eq "a1")
```

2. Rewriting the source query into target subqueries over the RDF or XML sources, by applying the query rewriting algorithm: RDQL2RDQL or RDQL2PXQuery (once for each source), which utilizes mapping information stored in the mapping table of Figure 4. The output Q_{out} of a query rewriting in algorithm is a triple of the form $(P_{Q_{out}^S}, P_{Q_{out}^W}, C_{Q_{out}})$ for the RDF source or $(V_{Q_{out}^R}, V_{Q_{out}^W}, C_{Q_{out}})$ for the XML source. From Q_{out}, we can compose the target query that is executable over the local source. Below is the result of this step for Example 1.

For the local RDF source R:
  $P_{Q_{out}^S}=${Book.booktitle}, $P_{Q_{out}^W}=${Book, Book.booktitle}, $C_{Q_{out}}=${}.
  The target RDF query is: SELECT ?booktitle
                              WHERE (?book, <lo:booktitle>, ?booktitle)

For the local XML source S_1:
  $V_{Q_{out}^R}=${/books/book/@booktitle}, $V_{Q_{out}^W}=${/books/book, /books/book/@booktitle,
  /books/book/author, /books/book/author/@name},
  $C_{Q_{out}}=${/books/book/author/@name, =, "a1"}.
  The target XML query is: for $book in doc("books.xml")/books/book
                              where $book/author/@name = "a1"
                              return $book/@booktitle

For the local XML source S_2:
  $V_{Q_{out}^R}=${/writers/writer/article/@title}, $V_{Q_{out}^W}=${/writers/writer/article,
  /writers/writer/article/@title, /writers/writer, /writers/writer/@fullname},
  $C_{Q_{out}}=${/writers/writer/@fullname, =, "a1"}.
  The target XML query is: for $writer in doc("writers.xml")/writers/writer
                              where $writer/@fullname = "a1"
                              return $writer/article/@title

3. Building an answer to the source query (on the global ontology G) by assembling the fragment results returned from local sources. We need to not only *union* the fragments (returned from different sources) while removing identical records, but also *join* the records based on some common key attribute. In addition, *null* values will be filled into the records that just partially cover queried attributes. The result of an RDQL query is a table containing URIs or string constants corresponding to the path expressions in the SELECT clause. For example, the answer to the query of Example 1 is a table containing a single tuple ("b1"), which is the union of results from S_1 and S_2. The record ("b3") returned from R is filtered out since the target query over R loses the query constraints in query rewriting, caused by the partial mappings from G to R (i.e., R has no correspondence for the class Author in G).

5.3 Query Answering in Hybrid P2P Mode

We only focus on the case of translating a source query in PXQuery from a peer to all the other peers, since the translation of a source RDQL query is similar to what is done in data integration mode (except for the transitive mappings). Query answering in hybrid P2P mode includes the following steps.

  1. Analyzing the source PXQuery query to convert it from a string to a triple $Q_{in} : (V_{Q_{in}^R}, V_{Q_{in}^W}, C_{Q_{in}})$.

Example 2. To "list all the publications", the user poses a query (over the local source S_1) as shown below on the left hand side. The resulting Q_{in} components are listed on the right hand side.

```
for $book in doc("books.xml")/books/book        V_{Q_{in}^R}={/books/book}
return $book                                     V_{Q_{in}^W}={}, C_{Q_{in}}={}
```

  2. Rewriting the source query into a target query over all the other connected RDF or XML sources, by utilizing the query rewriting algorithm: PXQuery2RDQL or PXQuery2PXQuery (once for each source) and the transitive mappings between the original data source and the target data source. The output of the query rewriting algorithm is a triple $Q_{out} : (V_{Q_{out}^R}, V_{Q_{out}^W}, C_{Q_{out}})$ for the target XML data source or $(P_{Q_{out}^S}, P_{Q_{out}^W}, C_{Q_{out}})$ for the target RDF data source.

An XML query must take into account the document structure of the XML source. The answer to an XML query is returned as a set of subtrees, each of which is rooted from one of the queried nodes (i.e., vertices in V_{Q^R}). For instance, the answer to the XML query in Example 2 is the subtree rooted from book in S_1 (see Figure 1). Therefore, the query rewriting algorithm also outputs a tree T with its children being the resulting subtrees of the answer. The result of this step by following Example 2 is shown below.

For the local RDF source R:
  $P_{Q_{out}^S}$={Book}, $P_{Q_{out}^W}$={}, $C_{Q_{out}}$={}.
  The target RDF query is:
    SELECT ?book, ?title
    WHERE (?book, <lo:booktitle>, ?title)

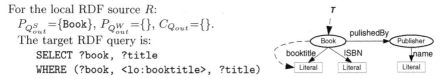

For the local XML source S_2:

$V_{Q_{out}^R} = \{/\texttt{writers/writer/article}\}, V_{Q_{out}^W} = \{\}, C_{Q_{out}} = \{\}.$

The target XML query is:

```
for $writer in doc("writers.xml")/writers/writer
    for $article in $writer/article
return
    <book booktitle="{$article/@title}">
        <author name="{$writer/@fullname}"/>
    </book>
```

3. Building an answer to the source query (against the original data source) by computing the union of the local answer (returned from the original queried peer) and the remote answers (returned from remote peers). To construct the remote answers, different methods are used for queries that target XML sources versus queries that target RDF sources. In the former case, because RDQL cannot represent document structure, the remote answer is built by organizing (based on the structure specified by T) the instances returned from executing the target RDQL query. Whereas in the latter case, the remote answer is formed by simply executing the target PXQuery query that already represents the same structure as specified by T. For Example 2, the final answer to the source query is shown below, where the three resulting lines come from the local sources S_1, S_2, and R, respectively.

```
<book booktitle="b1"> <author name="a1"> </book>
<book booktitle="b2"> <author name="a2"> <author name="a3"> </book>
<book booktitle="b4"> </book>
```

6   Conclusions and Future Work

In this paper, we propose a P2P schema-based data management framework called PEPSINT. This framework aims to semantically integrate distributed heterogeneous XML and RDF data sources. We discuss the construction of the architecture, maintenance of mappings, and query processing in PEPSINT. In particular, semantic integration is implemented at schema-level through the schema matching process and at instance-level through the query answering process. A key aspect in these two processes is the preservation of domain and document structure, which is realized by extending the RDF metadata space and providing a set of query rewriting algorithms. Because of this preservation, the user query can be correctly propagated across the heterogeneous XML and RDF data sources in PEPSINT, so that information access within the network is transparent to the user.

As for future work, we will: (1) Develop a proof of correctness for the query process. (2) Design and implement a semantic web application (e.g., for bibliographic data exchange) in PEPSINT to validate and evaluate the system. (3) Do a performance comparison of PEPSINT with other P2P data management systems.

References

1. B. Amann, C. Beeri, I. Fundulaki, and M. Scholl. Ontology-Based Integration of XML Web Resources. In *Proceedings of the 1st International Semantic Web Conference (ISWC 2002)*, pages 117–131, 2002.
2. B. Amann, I. Fundulaki, M. Scholl, C. Beeri, and A. Vercoustre. Mapping XML Fragments to Community Web Ontologies. In *Proceedings of the 4th International Workshop on the Web and Databases (WebDB 2001)*, pages 97–102, 2001.
3. M. Arenas, V. Kantere, A. Kementsietsidis, I. Kiringa, R. J. Miller, and J. My-lopoulos. The Hyperion Project: From Data Integration to Data Coordination. *SIGMOD Record*, 32(3):53–38, 2003.
4. S. Bergamaschi, F. Guerra, and M. Vincini. A Peer-to-Peer Information System for the Semantic Web. In *Proceedings of the International Workshop on Agents and Peer-to-Peer Computing (AP2PC2003)*, July 2003.
5. S. Boag, D. Chamberlin, M. F. Fernández, J. R. D. Florescu, and J. Siméon. XQuery 1.0: An XML Query Language. http://www.w3.org/TR/xquery, W3C Working Draft, August 2003.
6. S. D. Camillo, C. A. Heuser, and R. S. Mello. Querying Heterogeneous XML Sources through a Conceptual Schema. In *Proceedings of the 22nd International Conference on Conceptual Modeling (ER2003)*, pages 186–199, 2003.
7. Y. Chen and P. Revesz. CXQuery: A Novel XML Query Language. In *Proceedings of International Conference on Advances in Infrastructure for Electronic Business, Science, and Medicine on the Internet (SSGRR 2002w)*, 2002.
8. I. F. Cruz and H. Xiao. Using a Layered Approach for Interoperability on the Semantic Web. In *Fourth International Conference on Web Information Systems Engineering (WISE'03)*, pages 221–232, Roma, Italy, December 2003.
9. I. F. Cruz, H. Xiao, and F. Hsu. An Ontology-based Framework for Semantic Inter-operability between XML Sources. In *Eighth International Database Engineering & Applications Symposium (IDEAS 2004)*, July 2004. (To appear).
10. L. Gong. JXTA: A Network Programming Environment. *IEEE Internet Computing*, 5(3):88–95, May 2001.
11. A. Y. Halevy, Z. G. Ives, P. Mork, and I. Tatarinov. Piazza: Data Management Infrastructure for Semantic Web Applications. In *Proceedings of the 12th International World Wide Web Conference (WWW2003)*, pages 556–567, 2003.
12. HP Labs. RDQL - RDF Data Query Language. http://www.hpl.hp.com/semweb/rdql.htm.
13. M. Lenzerini. Data Integration: A Theoretical Perspective. In *Proceedings of the 21st ACM SIGACT-SIGMOD-SIGART Symposium on Principles of Database Systems (PODS 2002)*, pages 233–246, Madison, Wisconsin, June 2002. ACM.
14. G. Moro, A. M. Ouksel, and C. Sartori. Agents and Peer-to-Peer Computing: A Promising Combination of Paradigms. In *Proceedings of the 1st International Workshop of Agents and Peer-to-Peer Computing (AP2PC2002)*, pages 1–14, 2002.
15. W. Nejdl, B. Wolf, C. Qu, S. Decker, M. Sintek, A. Naeve, M. Nilsson, M. Palmér, and T. Risch. EDUTELLA: A P2P Networking Infrastructure Based on RDF. In *Proceedings of the 11th International World Wide Web Conference (WWW2002)*, 2002.
16. W. S. Ng, B. C. Ooi, K. Tan, and A. Zhou. PeerDB: A P2P-based System for Distributed Data Sharing. In *Proceedings of the 19th International Conference on Data Engineering (ICDE 2003)*, pages 633–644, 2003.
17. E. Rahm and P. A. Bernstein. A survey of approaches to automatic schema matching. *VLDB J.*, 10(4):334–350, 2001.

The SEWASIE Multi-agent System

Sonia Bergamaschi[1], Pablo R. Fillottrani[2], and Gionata Gelati[1]

[1] Dipartimento di Ingegneria dell'Informazione,
Università di Modena e Reggio Emilia,
Via Vignolese 905, 41100 Modena, Italy
sonia.bergamaschi@unimo.it
jonathan.gelati@unimore.it
[2] Faculty of Computer Science,
Free University of Bozen/Bolzano,
Piazza Domenicani 3, 39100 Bozen/Bolzano, Italy
fillottrani@inf.unibz.it

Abstract. Data integration, in the context of the web, faces new problems, due in particular to the heterogeneity of sources, to the fragmentation of the information and to the absence of a unique way to structure, and view information. In such areas, the traditional paradigms on which database foundations are based (i.e. client/server architecture, few sources containing large information) have to be overcome by new architectures. In this paper we propose a layered P2P architecture for mediator systems. Peers are information nodes which are coordinated by a multi-agent system in order to allow distributed query processing.

1 Introduction

The advancing of the Internet has opened the access to an overwhelming amount of data. While users can benefit of a vast information, data have an heterogeneous format and are sparsed over different places, making the search for data a costly operation. Integration of heterogeneous information in the context of the Internet becomes a key activity to enable a more organized and semantically meaningful access to data sources. If we look at the Internet as a P2P data-sharing system where peers are data sources, the challenge is twofold. First, peers presents information according to their particular view of the matter, i.e. each of them assumes a specific ontology. Second, data sources are usually isolated, i.e. they do not share any topological information concerning the content or the structure of other sources. The classical approach to solve these issues is provided by mediator systems which aim at creating a unified virtual view of the underlying data so as to hide the heterogeneity and distribution of data and give users a coherent access to the integrated information. Traditional solutions focus on the creation of one mediator system to integrate diverse data sources [1,2,3,4]. Our view is that next generation information systems will include a *network of mediator systems*, where mediators are not isolated any longer and are organized so that to share and map their ontologies. We propose here to use a P2P system

G. Moro, S. Bergamaschi, and K. Aberer (Eds.): AP2PC 2004, LNAI 3601, pp. 120–131, 2005.

where peers are mediator systems and are supported by a multi-agent system so as to propose users a mapped knowledge of the underlying ontologies. The multi-agent system organizes the peer network so as peer ontologies are shared and mapped. While single peers independently carry out their own integration activities, they exchange knowledge with agents which provide a coherent access to the underlying peer network. This defines two layers in the system: at local level, peers maintain an integrated view of local sources, at network level agents maintain mappings among the different peers. The result is the definition of a new type of mediator system intended to operate in web economies, called the SEWASIE system.

In section 2 we present the SEWASIE architecture and we explain how integration of information and query management work in section 3. The discussion of related work is presented in section 4 and final remarks in section 5.

2   SEWASIE Architecture

In this section both the functional and deployment architecture of the SEWASIE system are described.

2.1   The Functional Architecture

The SEWASIE architecture [5,6] is composed by a network of mediator systems and a set of agents to support users querying the underlying peers as a unique transparent data source (see Figure 1).

The mediator systems are the peers of the SEWASIE system and we call them *SEWASIE Information Nodes* (SINodes). SINodes are mediator-based systems [7], each including a *Virtual Data Store*, an *Ontology Builder*, and a *Query Manager*. A Virtual Data Store represents a virtual view of the overall information managed within an SINode and consists of the managed information sources, wrappers, and a metadata repository. The managed information sources are heterogeneous collections of structured, semi-structured, or unstructured data, e.g. relational databases, XML/HTML or text documents and are accessible by means of wrappers which are intended to translates to and from local access languages. There is one wrapper linked to each information source. According to the metadata provided by the wrappers, the Ontology Builder performs semantic enrichment processes in order to create and maintain the current ontology which is made up of the *Global Virtual View* (in short GVV), of the managed sources and the mapping description between the GVV itself and the integrated sources. Ontologies are built on a logical layer based on existing W3C standard. The Metadata Repository holds the ontology (GVV) and the knowledge required to establish semantic inter-relationships between the SINode itself and the neighbouring ones.

In order to support the network of peers in offering an integrated view over their ontologies, a set of agents has been defined. These agents cover functionalities required to keep knowledge of the topology of the system as well as the

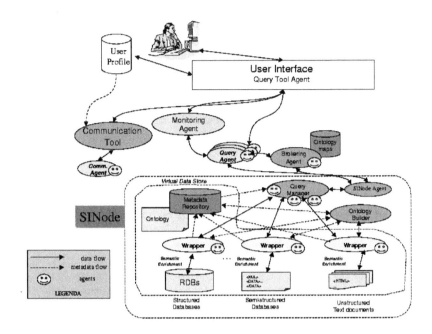

Fig. 1. The SEWASIE system architecture

semantical mappings that can be established among peers. The topology of the system is used to know which SINodes participate to the SEWASIE network, whether they are available in a certain moment to solve queries posed to the system or request to update the ontology. The semantical mappings are exploited in the query processing phase, where a query may involve more SINodes. More precisely, SEWASIE agents have four basic types: Brokering Agents, Query Agents, Monitoring Agents and Communication Agents.

Brokering Agents are the peers responsible for maintaining a view of the knowledge handled by the network. This view is maintained in *ontology mappings*, that are composed by the information on the specific content of the SINodes which are registered by the brokering agent, and also by the information on the content of other brokering agents. Thus, brokering agents must provide means to publish the locally held information within the network.

Query Agents are the carriers of the user query from the user interface to the SINodes, and have the task of solving a query by interacting with the brokering agent network. Currently, query agents are able to interact with one brokering agent. Future versions of the system will include query agents dealing with more brokering agents. Once a brokering agent is contacted, it informs the query agent which SINodes under its control contain relevant information for the query. Then, the query agent asks the involved SINodes for collecting partial results. Also, it decides whether to continue the search with the other brokering agents. Once this process is over, all partial results are fused into a final answer to be delivered to the user.

A *Query Tool Agent* is part of the SEWASIE user interface. It includes a query tool that guides the user in composing queries. A query tool agent is responsible for contacting brokering agents in order to get ontologies and is also responsible to manage the set of query agents required to solve users' queries.

Monitoring Agents and *Communication Agents* provide user-oriented services. Monitoring agents are responsible for monitoring information sources according to user interests which are defined in *monitoring profiles*. Each monitoring agent is assigned a specific topic of interest chosen by one user. Each monitoring agent contains an internal ontology, i.e. a *domain model*, which is linked to brokering agents ontologies. Agents of this type regularly set up query agents to query the SEWASIE network, filter the results, and fill *monitoring repositories* with observed documents.

A communication agent supports negotiation between one user and other parties present in the SEWASIE network (usually parties that have exposed an SINode). Any query including contact information sets the context to launch the communication. Several types of communication agents can be created for one communication each helping find and contact potential business partner, asking for initial offers, and ranking them. The human negotiator can then decide and choose the best offer to begin negotiating, with support from the communication tool. This latter maintains four types of agents, that can act in the different phases of the negotiation [8]. The Initiation Agent tries to establish contacts with potential partners according to the a user's preferences. The Filtering and Ranking Agent maintains the overview of the negotiation process (containing several parallel negotiations) and provides support for decision making by calculating the scores of received offers and ranking them. The main task of the Resource Management Agent is to notify the user when resources lack. The Negotiation Agent can act when the negotiation achieves a well structured and defined state. In this case the Negotiation Agent tries to provide some offers depending on user defined preferences and negotiation strategy.

2.2   System Deployment

So far we have presented the functional architecture of the system. We now want to shortly describe its deploying architecture. The SEWASIE system is intended to operate in networked environments where heterogeneity and distribution of information is a reality. Peers, i.e. SINodes, expose their GVVs on the network and software agents act as a glue among the different peers. Peers are recognised as being part of the SEWASIE system as long as they register their GVVs by a brokering agent. From a deployment view point, what is distributed is the multi-agent system. As the scope of the SEWASIE project is to focus on the application of software agents and not in providing a general toolkit for building multi-agent systems, the choice was to use existing tools. The key features we were looking for were:

- a high-level language, in order to focus on application programming;
- portability, in order to allow for multiple platforms to become part of the SEWASIE system in a transparent way;

- FIPA compliance, in order to be aligned with the current standards for agent technology;
- support and maintenance, in order to meet deployment needs.

Currently, the number of alternative agent toolkits is quite good [9–17]. Our choice felt on the Java Agent DEvelopment (JADE) developed by TILab [16]. JADE is currently one of the most evolving toolkits and is an open source. JADE is written in Java [18] and exploits Java RMI [19] for managing software distribution in the environment.

A JADE multi-agent system (or *platform*) is a logical space that can be distributed over diverse physical hosts. Each host participating to the platform has its own Java Virtual Machine (JVM) running. Each JVM is an agent container, i.e. a runtime environment that allows agents to concurrently execute. In order to boot the platform, a main container has to be created. The main container hosts the services necessary to support agents' life cycle, migration and communication. Containers eventually residing on remote hosts can be added to the platform at runtime. No matter where containers are located, the agent platform is seen as a uniform logical space, where all containers can be reached simply knowing their name. Recently, JADE introduced the support for security as an extension of the Java security model and in particular of the JAAS interface [20]. Besides the JADE security extension, we have exploited tunneling techniques in order to address security issues related to network configurations. This has been necessary to deploy the system in firewalled environments.

Tunnelling is a popular technique which permits to expose services on the standard port of a web server. Client applications can then reach the service by executing an HTTP request. The web server will redirect the request to the particular service addressed. Responses are sent following the backward path.

Our deployment architecture foresees therefore that each host activates a web server. The web server acts as a gate to the network environment. Messages and objects to and from an agent container belonging to the platform are HTTP requests going through the web server. This is made possible because Jade manages remote objects and remote calls using Java RMI. When an RMI server is activated, a registry to keep track of all (possibly remote) objects registered is initiated which listen to incoming requests on a given port number. An RMI client can call this service in order to remotely connect and use objects. The web server can make accessible the RMI server in two ways either through a CGI script or by means of a servlet activated in an application server. While the CGI script requires less infrastructural component, the servlet offers higher performance. This represents a tradeoff.

3   Integrating Knowledge and Querying

In this section we describe how the SEWASIE system comes into being and which are the mechanisms required to maintain the system work, restricting our attention on how information is integrated and how query processing happens.

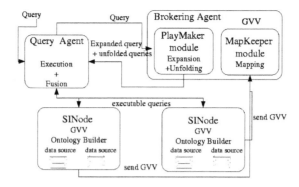

Fig. 2. SEWASIE integration and querying

3.1 Integrating Knowledge: SINodes and Brokering Agents

As we have seen in the previous section, data are actually stored in local data sources managed by an SINode. Therefore, the first step towards integrating data has to be undertaken at the level of SINodes (see lower part of Figure 2). In order to create and maintain a global view of its information sources, SINodes require an *Ontology Builder*. The Ontology Builder is the collective name of a set of functionalities which support the creation and maintenance of the GVV of an SINode. It helps synthesize ontologies and merging them into a GVV. The building process begins with the creation of a Common Thesaurus of the information provided by wrappers. The thesaurus is obtained by annotating the schema of the sources and by inferring the terminological intensional and extensional relationships based on such annotations and describing intra-schema knowledge about classes and attributes of each source schema. Based on such information and on designer supplied relationships capturing specific domain knowledge, the Ontology Builder module performs semi-automatic inter-schema analysis by exploiting lexicon derived relationships (which are based on processes like synonym identification or generalisation-specialisation relations) and by inferring new relationships. All these relationships are considered in the subsequent phase of ontology building, which performs hierarchical clustering and supports the emergence of a number of global classes representative of all the classes coming from the sources (the GVVs) and of a set of mappings between the GVV and the local sources. A full description of the integration steps can be found in [4].

3.2 Glueing Peers

SINodes constitute the network of peers. By themselves peers are not aware of the presence of other peers and have no capability to integrate external knowledge. What activates the network are brokering agents. In order to be part of the SEWASIE system, i.e. share the ontology with others, an SINode must register with at least one brokering agent. This happens as follows. An SINode wishing to become part of a SEWASIE system links to the SEWASIE agent platform. This is done by subscribing a newly created agent container and activating a new SINode agent, responsible for interfacing the SINode with the SEWASIE agent platform. These actions are usally undertaken by the SINode administrator. The knowledge about the host (and port) on which the SEWASIE platform is running is required. Alternative ways to interface an SINode with the SEWASIE agent platform exist. For instance, we may use Web Services which interact with the JADE platform as an external wrapped application[1]. The solution with an SINode agent is the most flexible one as it makes available all the agent features for interacting with an agent platform and other agents. In the following we will refer to an SINode as an agent. The same mechanisms can be viewed as if a Web Service was instead running.

Upon the arrival of some particular event (most likely the GVV has been created and can now be exported to the network), the SINode starts the advertising phase, where it asks to the currently executing brokering agents to integrate its GVV. Figure 3 depicts the AUML sequence diagram of the interaction using templates [21]. The full list of available brokering agents is retrieved by querying the *Directory Facilitator* service (DF). The DF is the standard JADE yellow-pages service: agents can advertise to the DF their capabilites and keep updated the information about their status. The list of brokering agents may be further filtered according to selective parameters, such as the number of GVVs already integrated by a brokering agent or to its workload. The SINode agent will then contact the selected brokering agents requesting the integration of its GVV. Brokering agents can decide whether to satify the request or not. This phase ends successfully if at least one brokering agent accepts to integrate the view. If unsuccessful, the SINode agent may have a later try. The accepting brokering agents can now integrate the sent GVV. The module responsible for this operation is called *Map Keeper* (see Figure 2). Integrating means building the mappings among the diverse GVVs which have been collected by the brokering agent. This activity is similar to the one carried out by SINodes when integrating knowledge from different data sources. What is different now is that mappings can be established more easily as each source schema is already represented in a standard format and its semantics is expressed. The process configures thus as automatic with few or no intervention from an ontology designer. A map keeper keeps its own GVV built on top of the collected GVVs coming from SINodes.

In dynamic settings not only data stored in the sources can change but schemas can evolve over time. Mechanisms on how changes in data source schemas can be reflected in an existing GVV are under investigation [22,23].

3.3   Querying

The overall integration process results in brokering agents each producing a GVV of the schema of the managed SINodes. An SINode can be managed by more brokering agents. Users have access to the GVV of the brokering agents

---

[1] JADE provides a specific package which handles the details of in-process JADE management.

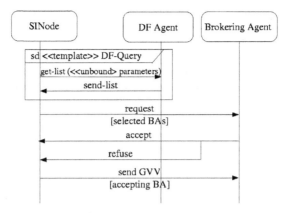

Fig. 3. The AUML diagram of the interaction betweeen an SINode and brokering agents

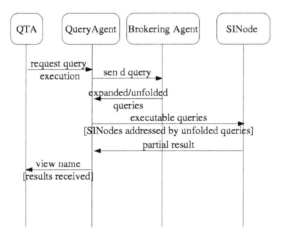

Fig. 4. The AUML diagram of the interaction occuring during the query solving phase

by means of the query tool agent interface. While navigating these views they may also pose query. We now describe how query management is carried out in the SEWASIE system with reference to a simplified architecture with a single brokering agent. Figure 4 depicts the AUML sequence diagram of the interaction we will describe. More details are reported in [24]. The Query Tool Agent is the one that receives the query from a user. It starts a query management phase by sending the query to a query agent.

In order to execute the query, a query agent must know which SINodes have to be contacted. This information is asked the brokering agent (see Figure 2). Given the query, a brokering agent is capable of decomposing it using its GVV so as to produce queries that are executable by SINodes. In general, a query posed by a user maps into a set of queries to be sent to SINodes and thus a brokering agent further computes the query required to fuse the partial results.

Decomposing the original query into executable ones according to the mapping of the GVV must satisfy requirements related to the correctness and completeness of the answer. In the context of the integrated schemas, correctness translates into assuring that the constraints enforced on the brokering agent GVV are respected when decomposing the query, while completeness translates into assuring that both the required join operations to fuse the final answer are performed and the necessary filtering conditions are appropriately applied for each executed query. A brokering agent performs thus two steps: (a) query expansion, which expands the query taking into consideration the integrity constraints of the global schema and (b) query unfolding which decomposes a query into queries executable by local sources so as they can be coherently fused to get the final answer.

As shown in Figure 2, the overall outcome is an expanded query and a set of unfolded queries which comprise queries executable by SINodes and queries to coherently fuse their results. The obtained set of queries identifies the set of SINodes addressed by the original query. All these data are passed back to the requesting query agent. At this stage, the query agent is responsible for supervisioning the execution of the queries. It basically sends the executable part of the unfolded queries to each SINode, collects the partial answers, applies the residual filtering conditions and resolution functions and finally fuse the final answer solving the expanded query. The result is returned into the form of a view name which can be queried by the query tool agent to visualize results to the requesting user.

4   Related Work

Several agent-based information retrieval systems are known. In order to compare to similar systems, we now emphasize SEWASIE main characteristics:

- two-level data integration scheme: strongly tied local nodes are integrated into SINodes; BAs provide globally integrated ontologies by means of weaker mappings.
- query management: query building assisted by a query tool, query rewriting in the two levels of data integration following local ontologies using sound and complete algorithms.
- additional tools: negotiation and monitoring tools integrated in the same agent architecture.

Altogether these points make the SEWASIE system unique among the agent-based information retrieval systems.

Some systems are strong on data integration. CARROT II [25] is an agent-based architecture for distributed information retrieval and document collection management. It consists of an arbitrary number of agents providing search services over local document collections or information sources. They contain metadata describing their local document store which are sent to other agents that act as brokers. Like in SEWASIE, these metadata have an unstructured form, without a central control. But there are anyway several differences with the SEWASIE architecture. First, data integration is done in only one level. In this sense, CARROT II agents play the role of a brokering agent and an SINode at the same time. Second, there is no support for the user in creating the query. Metadata information is not reflected in the process of query building. Finally, the most important difference is that agents in this system only produce a routing of the query to relevant information sources, no query rewriting is done in this step. In SEWASIE the query is reformulated following brokering agent's ontology before asking SINodes, which contain the information sources. Several other information retrieval systems are known with routing agents, like HARVEST [26], CORI [27] and InfoSleuth [28].

Other systems, like TSIMMIS [29], include some rewriting rules against predefined query patterns. There are several steps of query processing also in the MISSION project [30]. In these cases, data integration technology is not present, or in TSIMMIS limited to automatic generation of wrappers [31] and mediators [32] from web pages. In SEWASIE, the data integration techniques [4] adopted by SINodes apply not only to unstructured, or semi-structured data sources, but also to relational databases.

5   Conclusions and Future Work

In this paper, we have provided a general description of the SEWASIE multi-agent architecture. We have shown the different types of agents and how they are organized. While tackling architectural issues, we have made some observation on the deployment architecture of the SEWASIE system. We have then described how the system works for integrating and querying data. As future work we will study how to extend the presented model to a setting where more brokering agents are active. During the integration process, SINodes may then register with multiple SINodes and SINodes may exchange information concerning their mappings. As for solving query, query agents can contact more brokering agents in order to obtain a decomposed query.

Acknowledgements

This work is supported in part by the 5th Framework IST programme of the European Community through project SEWASIE within the Semantic Web Action Line.

References

1. Garcia-Molina, H., Hammer, J., Ireland, K., Papakonstantinou, Y., rey, J., Jennifer, U.: Integrating and accessing heterogeneous information sources in tsimmis (1995)
2. Chawathe, S., Garcia-Molina, H., Hammer, J., Ireland, K., Papakonstantinou, Y., Ullman, J.D., Widom, J.: The TSIMMIS project: Integration of heterogeneous information sources. In: 16th Meeting of the Information Processing Society of Japan, Tokyo, Japan (1994) 7–18

3. Kirk, T., Levy, A.Y., Sagiv, Y., Srivastava, D.: The Information Manifold. In Knoblock, C., Levy, A., eds.: Information Gathering from Heterogeneous, Distributed Environments, Stanford University, Stanford, California (1995)
4. Bergamaschi, S., Castano, S., Beneventano, D., Vincini, M.: Retrieving and integrating data from multiple sources: the MOMIS approach. Data and Knowledge Engineering **36** (2001) 215–249
5. Bergamaschi, S.: Global architecture of the SEWASIE system. SEWASIE Deliverable D1.3a (2003)
6. Bergamaschi, S., Guerra, F., Vincini, M.: A peer-to-peer information system for the semantic web. In: 2nd. International Workshop on Agents and Peer-to-Peer Computing, held in AAMAS 2003 International Conference on Autonomous Agents and MultiAgent Systems. (2003)
7. Wiederhold, G.: Mediators in the architecture of future information systems. IEEE Computer **25** (1992) 38–49
8. Schoop, M., Rehman, M.U., Jertila, A.: Specification of agent technology for negotiation support. SEWASIE Deliverable D5.4 (2004)
9. Tryllian: Agent Development Kit. `www.tryllian.com/technology/product1.html` (1999)
10. Aisland: Aisland Project. `aisland.jxta.org` (2001)
11. Labs, F.: April Agent Platform. `www.nar.fujitsulabs.com/aap` (1997)
12. Comtec: Comtec Agent Platform. `ias.comtec.co.jp/ap` (2000)
13. FIPA-OS: FIPA-OS. `fipa-os.sourceforge.net` (1997)
14. Grasshopper: Grasshopper. `www.grasshopper.de` (1999)
15. Software, A.: JACK intelligent agents. `www.agent-software.com` (1998)
16. (TILab), T.I.L.: JADE. `jade.cselt.it` (2000)
17. Agent, J.: JAS API. `www.java-agent.org` (2000)
18. Microsystem, S.: http://java.sun.com/ (2003)
19. Microsystem, S.: http://java.sun.com/j2se/1.4.2/docs/guide/rmi/ (2003)
20. Microsystem, S.: http://java.sun.com/products/jaas/ (2003)
21. Modeling, F.T.: Fipa modeling: Interaction diagrams (2003)
22. Fergnani, A.: Ontology dynamics for semantic web: the momis approach (2002)
23. Beneventano, D., Bergamaschi, S., Guerra, F., Vincini, M.: Synthesizing an integrated ontology. IEEE Internet Computing Magazine **7** (2003) 42–51
24. Beneventano, D., Lenzerini, M., Majkić, Z., Mandreoli, F.: Techniques for query reformulation, query merging, and information reconciliation - part a. SEWASIE Deliverable D3.2a (2003)
25. Klusch, M., Ossowski, S., Shehory, O., eds.: Integrating Distributed Information Sources with CARROT II. In Klusch, M., Ossowski, S., Shehory, O., eds.: Cooperative Information Agents VI, 6th International Workshop, CIA 2002, Madrid, Spain, September 18-20, 2002, Proceedings. Volume 2446 of Lecture Notes in Computer Science., Springer (2002)
26. Bowman, C.M., Danzig, P., Hardy, D.R., Manber, U., Schwartz, M.F.: The Harvest information discovery and access system. Computer Networks and ISDN Systems **28** (1995) 119–125
27. Callan, J.P., Lu, Z., Croft, W.B.: Searching distributed collections with inference networks. In Fox, E.A., Ingwersen, P., Fidel, R., eds.: SIGIR'95, Proceedings of the 18th Annual International ACM SIGIR Conference on Research and Development in Information Retrieval. Seattle, Washington, USA, July 9-13, 1995 (Special Issue of the SIGIR Forum), ACM Press (1995) 21–28
28. Woelk, D., Tomlinson, C.: Infosleuth: Networked exploitation of information using semantic agents. In: COMPCON Conference. (1995)

29. Garcia-Molina, H., Papakonstantinou, Y., Quass, D., Rajaraman, A., Sagiv, Y., Ullman, J.D., Vassalos, V., Widom, J.: The TSIMMIS approach to mediation: Data models and languages. Journal of Intelligent Information Systems **8** (1997) 117–132

30. McClean, S.I., Karali, I., Scotney, B.W., Greer, K., Kapos, G.D., Hong, J., Bell, D.A., Hatzopoulos, M.: Agents for querying distributed statistical databases over the internet. International Journal on Artificial Intelligence Tools **11** (2002) 63–94

31. Hammer, J., McHugh, J., Garcia-Molina, H.: Semistructured data: The tsimmis experience. In: Proceedings of the First East-European Symposium on Advances in Databases and Information Systems (ADBIS'97), St.-Petersburg, September 2-5, 1997. Volume 1: Regular Papers, Nevsky Dialect (1997) 1–8

32. Papakonstantinou, Y., Garcia-Molina, H., Widom, J.: Object exchange across heterogeneous information sources. In Yu, P.S., Chen, A.L.P., eds.: Proceedings of the Eleventh International Conference on Data Engineering, March 6-10, 1995, Taipei, Taiwan, IEEE Computer Society (1995) 251–260

Service Discovery on Dynamic Peer-to-Peer Networks Using Mobile Agents

Evan A. Sultanik and William C. Regli

Department of Computer Science, College of Engineering,
Drexel University, Philadelphia, PA 19104-2875
{eas28, regli}@drexel.edu

Abstract. Service discovery and location introduces numerous challenges for multi-agent planning in dynamic, real-world domains. Specifically, on non-fault tolerant, peer-to-peer and ad hoc wireless networks, services and agents may become unavailable due to network partitioning, traffic congestion, or attack. Such disruptions might prohibit the ability of agents to find the services needed to execute a plan, possibly threatening the survivability and stability of the overall agent system. This research introduces a method for service discovery and availability prediction based on random walks and demonstrates its applicability in the setting of peer-to-peer, wireless networks.

1 Introduction

Dynamic multi-agent planning has many shortcomings in real-world domains. Specifically, it is not clear how services can be discovered and located in dynamic non-fault-tolerant networks. Examples of such include multi-hop ad hoc wireless and dynamic peer-to-peer networks. The synthetic aircraft domain [1] affirms the problem of service discovery for multi-agent planning systems. A group of agent-driven helicopters are deployed on a battlefield; one helicopter "disappears." In this case "disappearance" might mean "over a ridge," "out of communication range," or "destroyed." How do the remaining agents decide if a node, or service on that node, has become unavailable and re-planning is required? Existing work assumes that such information is instantaneously announced to all nodes and agents, a process which fails to take into account the realities of information propagation on peer-to-peer and wireless networks. In the synthetic aircraft example, knowledge and intervention of a human agent is required to alert the agent system that the helicopter has disappeared. Furthermore, recent research has shown that no fixed memory deterministic algorithm can locate a service in a network in a fixed amount of time [2].

Our work addresses the problem of how agents can achieve global state awareness in peer-to-peer networks, with a specific focus on multi-hop, ad hoc wireless networks. In this context, "global state awareness" includes (but is not limited to) information about the location and capabilities of services on the dynamic network (i.e. mobile agents and web services). We propose a fixed-memory randomized method for approximating the location of a service in a dynamic network

G. Moro, S. Bergamaschi, and K. Aberer (Eds.): AP2PC 2004, LNAI 3601, pp. 132–143, 2005.

with a probabilistic certainty in a fixed amount of time. We present a mathematical formulation of agent movement in the network and an algorithmic technique for approximating the parameters of the formulation.

2   Background

The nature of ad hoc and peer-to-peer networks inherently imposes the restriction of communication to one's "neighbors." Therefore, communication over multiple hops in the network is difficult without facilities such as ad hoc routing [3,4,5,6]. Recent research has focused on using mobile agents for routing in ad hoc networks [7,8,9] and the need for further investigation in this area has been established [10]. Most methods rely on using the frequency of agents walking the network visiting each host to predict the topology. However, far less work exists on discovering the current distribution of *services* throughout the network.

Mainly due to the proliferation of peer-to-peer file sharing technology, significant research has been made in the area of dynamic networks. Service discovery architectures exist for such networks [11,12], some specifically designed for mobile ad hoc networks [13], however these often require services to register themselves when available. Search techniques have also been applied to the problem of localization, such as Content-Addressable Networks (CANs) [14,15]. However, search-based approaches often assume that agents can arbitrarily migrate from any host in the network to another [16]. Agent-based methods for service discovery have also been proposed [17,18]. Furthermore, CANs and Distributed Hash Tables assume that the index and data elements are fixed; none of these approaches directly address the problem of locating *mobile* services.

In some domains mobile agents can be service providers. For example, a certificate authority might be required for secure communication between helicopters in the synthetic aircraft domain. This server could be encapsulated by a mobile agent capable of reasoning about the network [19]. The agent might then continuously migrate to portions of the network with low volatility. For instance, the agent might migrate to helicopters less likely to be removed from the network. To improve performance and minimize latency, heuristics for such migration might include proximity to the geographic center of the group or association with the centroid of the network topology graph. Therefore, a method for pro-actively tracking the location of services in dynamic networks is required.

3   Theoretical Approach

Deploying a set of service monitoring agents, $\mathcal{A}$, to randomly walk a set of hosts, $\mathcal{H}$, of a peer-to-peer network can provide an accurate on-line means of service discovery. The agents act like bees, working to "pollinate" the network with their knowledge of services' locations. As seen in Figure 1, each agent's walk is dictated by the underlying network topology. Furthermore, there is no guarantee that all agents visiting a host have encountered a service thus far on their walks.

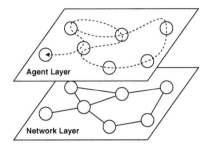

Fig. 1. An agent randomly walking a peer-to-peer network.

This approach has three important advantages over alternatives, such as naïve message passing and broadcast:

1. minimal network bandwidth is used, and bandwidth usage scales linearly; this is an important issue for resource-constrained mobile devices and large-scale peer-to-peer networks;
2. mobile code provides for synergy between networks of non-homogeneous service discovery architectures;
3. services need not register themselves; and
4. properties of random walks are relatively easy to mathematically model and likewise make inferences upon.

The remainder of this paper defines a mathematical model for the behavior of these agents, allowing for time-critical reasoning and probabilistic inferences to be made upon the system. We use properties of Markov chains to model the probability that an agent will visit a specific host, a binomial distribution to model the probability that an agent has seen a service, and another binomial distribution to develop an expected value for the number of agents that will visit a specific host in some amount of time.

3.1   Random Walking Mobile Agents

The agents' task environment, a dynamic peer-to-peer network, is stochastic, dynamic, and continuous; there exists a delay between actual topology changes and the propagation of knowledge of these changes throughout the network. The agents do not have a goal, per se; their sole purpose is to randomly walk the network gathering information. Agents' percepts are comprised solely of the set of services available at the current host, S_h, and the set of hosts neighboring the current host, $\{x \in \mathcal{H} | E_{h,x} > 0\}$ (where $E \subseteq \mathcal{H} \times \mathcal{H}$ is the set of edges in the topology and the notation $E_{x,y}$ denotes the weight of the edge from node x to node y). Edge weights represent transition probabilities between hosts in the network. For most networks these will be uniform. However, ad hoc wireless networks might correlate edge weights to link quality between hosts to avoid agent migration over unreliable links.

Agents' actions are comprised solely of hopping to a neighbor host from their current host. At each host agents query for services, storing these data in memory (along with a timestamp). The agents' itineraries are dictated by the network; successor hosts for migration are selected randomly from the set of available neighbor hosts in the network.

3.2   Predicting Agent Arrival at a Host

The frequency of agent visits can be predicted by developing a function, $F(N)$, for the probability that an agent $a \in \mathcal{A}$ with knowledge of a service $s \in \mathcal{S}$ will visit a specific host $h \in \mathcal{H}$ in a time interval t. N is a local state description represented by a tuple containing the following elements:

t - length of the time interval;
ν - probability an agent will be at host h;
η - number of instances of the service s;
$|\mathcal{H}|$ - cardinality of the set of hosts;
$|\mathcal{A}|$ - cardinality of the set of agents;
ℓ - average time needed for an agent to hop between neighbors; and
τ - maximum desired amount of time since an agent last saw the service.

We define $F(N)$ as a mapping from N to a real number probability:

$$F : (t, \nu, \eta, |\mathcal{H}|, |\mathcal{A}|, \ell, \tau) \mapsto [0, 1] \,. \tag{1}$$

$F(N)$ is therefore a probability distribution over the space N. We decompose $F(N)$ into three component distributions:

π_h - probability that an agent will visit host h at any given time;
$\hat{\nu}$ - an approximation of ν; and
$P(n \geq 1)$ - probability that an agent will see at least 1 instance of the service s in time τ.

These distributions need not be calculated a priori. A method for approximating π_h is given in §3.3 which is then used to develop $\hat{\nu}$ in §3.4. $P(n \geq 1)$ is then defined in §3.5. Finally, the approximation of $F(N)$ is constructed from these component distributions in §3.6.

Intuitively, the larger $|\mathcal{A}|$ and the smaller $|\mathcal{H}|$ & ℓ the more often agents will visit hosts. τ is simply meant to be a measure of the "age" of each agent's knowledge of the services. Since global time synchronization is a difficult problem in some domains, τ can also be replaced by a heuristic that approximates the age of the agent's data. For example, the number of hosts the agent has visited since it last saw an instance of the service could be used.

3.3   Mathematics of Random Walks

Random walks along graphs are essentially finite Markov chains, and both share many of the same properties. Gkantsidis, et al., experimentally showed that, when searching for items occurring frequently in a network, random walks perform better than flooding (for the same number of network messages) in certain cases [20]. In order to predict the frequency of randomly-walking agents visiting a specific host, though, we must first develop a probability that an agent will be on a specific host at any time.

The PageRank algorithm [21] determines the probability that a random web surfer will be on a given web page at any time. PageRank employs Markov chains to model random walks along the graph of the Internet. One can therefore use PageRank to determine the probability that an agent randomly walking a network will be visiting a specific host at any given time. The first eigenvector, π, of a graph's adjacency matrix, J, is fundamentally intertwined with the *stationarity* of the graph. The eigenvector π corresponds to the eigenvalue λ_1 such that $\pi J = \lambda_1 \pi$. PageRank exploits this fact and provides a means for approximating the primary right eigenvector of an adjacency matrix. Algorithm 1 is an adaptation of this algorithm for our domain.

Algorithm 1. AGENT-VISITATION-PROBABILITIES($J, d, iterations$)

Require: J is the adjacency matrix representation of the network, d is a real number damping factor in the range $[0, 1]$ (usually set to 0.85), *iterations* is the number of iterations to run, and all elements of π are initialized to $\frac{1}{|\mathcal{H}|}$.

Ensure: π, the primary right eigenvector of J, contains the probabilities that a random agent will be on any node

  for $i = 1$ to *iterations* **do**

    for $j = 1$ to $|\mathcal{H}|$ **do**

      $sum \leftarrow 0$

      for $k = 1$ to $|\mathcal{H}|$ **do**

        if $J_{k,j} > 0$ **then**

          $links \leftarrow |\{x \mid (1 \leq x \leq |\mathcal{H}|) \wedge (J_{k,x} > 0)\}|$

          if $links > 0$ **then**

            $sum \leftarrow sum + \pi_k \div links$

      $\pi_j \leftarrow (1 - d) + d \cdot sum$

π_h is then the probability that an agent $a \in \mathcal{A}$ is on a specific host $h \in \mathcal{H}$ at any given time. Mathematically, π can also be represented as follows:

$$\pi = \text{EIGENVECTOR}_1(dJ^T).\qquad(2)$$

3.4   Approximating ν

An element of the state description N, ν is defined as the probability that an agent will visit the host in time t. Therefore, by definition, one can use π to develop an estimator of ν:

$$\hat{\nu} = \pi_h, \ |\mathcal{A}| = 1.\qquad(3)$$

However, a binomial distribution must be used to define $\hat{\nu}$ if $|\mathcal{A}| > 1$:

$$\hat{\nu} = 1 - \left(1 - \left(1 - (1 - \pi_h)^{|\mathcal{A}|}\right)\right)^t$$

$$= 1 - (1 - \pi_h)^{t\,|\mathcal{A}|} . \tag{4}$$

In other words, $\hat{\nu}$ is 1.0 minus the probability that *none* of the $|\mathcal{A}|$ agents will visit in time t.

However, it is not sufficient to define the mapping of $F(N)$ solely based upon $\hat{\nu}$ because not all agents that visit a host have recent enough data about the service being located. Therefore, a function defining the probability that the random agent visiting host h will have a recent-enough[1] memory of service s is needed.

3.5   Accounting for τ

ν provides a prediction mechanism for the number of random-walking agents that will visit a host. However, ν does not take into account the fact that these agents may not have recently seen an instance of the service s. τ is an element of the state description that dictates the maximum amount of time since a service discovery agent has seen the service. Therefore, τ must be incorporated into the probability.

Let $H \subseteq \mathcal{H}$ be the set of hosts that an agent visits in time τ and $S \subseteq \mathcal{S}$ be the set of services an agent sees in time τ. The expected value for the number of hosts the agent will visit is given by:

$$\langle |H| \rangle = \left\lfloor \frac{\tau}{\ell} \right\rfloor . \tag{5}$$

It is assumed that, due to the mobility of the network and its associated random topology, the probability of a randomly-walking agent visiting a host with a service is normally distributed. This claim is empirically validated in §4.2. We can then say $\frac{\eta}{|\mathcal{H}|}$ is the probability that an instance of the service exists at a randomly-selected host. The probability that an agent walking the network will encounter an instance of s in time τ can then be modeled as a binomial distribution of $|H|$ trials:

$$P\left(n \,\|H\right) = \binom{|H|}{n} \left(\frac{\eta}{|\mathcal{H}|}\right)^n \left(1 - \frac{\eta}{|\mathcal{H}|}\right)^{|H|-n} , \tag{6}$$

where n is the number of instances of s discovered:

$$n = |\{x | x \in S \wedge x = s\}| . \tag{7}$$

---

[1] By "recent-enough" we mean "of age less than or equal to τ."

Summing (6) over all n where $n \geq 1$:

$$P(n \geq 1) = \sum_{i=1}^{|H|} \left(\binom{|H|}{i} \left(\frac{\eta}{|\mathcal{H}|} \right)^i \left(1 - \frac{\eta}{|\mathcal{H}|} \right)^{|H|-i} \right)$$

$$= 1 - \left(1 - \frac{\eta}{|\mathcal{H}|} \right)^{\lfloor \frac{\tau}{\ell} \rfloor}. \tag{8}$$

$P(n \geq 1) = P(\exists x, x \in S \wedge x = s)$ is the probability that an agent has seen at least one instance of the service s while walking the network in time τ.

3.6   Constructing $F(N)$

Given the probability that an agent will visit a host, ν, and the probability that a randomly walking agent will have seen an instance of the service, $P(n \geq 1)$, we can define the mapping of $F(N)$.

We assume the event that an agent has seen an instance of service s is independent of the agent visiting host h. Let A be the set of agents that visit h in time t. Using A, we can combine equations (4) & (8) and say,

$$F(N) = \begin{cases} 1.0, & A \neq \emptyset \\ 0, & A = \emptyset \end{cases}$$

$$= P(n \geq 1) \, \nu, \ t = 1. \tag{9}$$

Then, using another binomial distribution, we can define the mapping of $F(N)$ for all values of t:

$$F(N) = P(A \neq \emptyset)$$

$$= \sum_{i=1}^{t} \left(\binom{t}{i} (P(n \geq 1) \, \nu)^i \, (1 - P(n \geq 1) \, \nu)^{t-i} \right)$$

$$= 1 - \left(1 - \nu + \nu \left(1 - \frac{\eta}{|\mathcal{H}|} \right)^{\lfloor \frac{\tau}{\ell} \rfloor} \right)^t. \tag{10}$$

The function $F(N)$ is useful for predicting the number of randomly walking agents that have seen service s in time τ and will also visit host h in time t. Take the synthetic aircraft domain as an example; suppose an agent needs to locate a service in a fixed amount of time. If s does not exist, the agent will need to re-plan. If the service is only available from the helicopter that has disappeared, the agent will waste its time trying to look for the service. Using $F(N)$, the agent can predict if it will hear from any of the service discovery agents in time t. If $F(N)$ returns a low probability, the agent will know to immediately re-plan without waiting for any of the service-discovery agents to arrive.

4   Empirical Validation

4.1   Methodology

Network simulation is accomplished using the Macro Agent Transport Event-based Simulator (MATES) [22]. We model the network using the *exact connectivity* method for mobile ad hoc network graph generation [23]; connections are determined by the Euclidean distance between hosts. The hosts' movements are bounded by a 1200x1200 meter box, and each host has a radio range of 300 meters. At the beginning of each experiment, hosts are placed randomly in the box and given a random direction. Both agents and instances of the service s are randomly distributed among the hosts. Every iteration of the simulation:

1. Each host's direction is randomly chosen by either maintaining in its current heading (with probability 0.6), rotating 45° clockwise (with probability 0.2), or rotating 45° counter-clockwise (with probability 0.2);
2. hosts move forward one meter in their respective directions;
3. agents not currently in transit migrate from their current host to a randomly-selected neighbor host (as described in §3.1). Agent transit times are calculated with an inverse exponential relationship to the Euclidean distance between hosts. The average transit time for agents, ℓ, is 1 iteration;
4. every instance of s is treated as a mobile agent; each service migrates to a random neighbor host as described above; and
5. every t iterations, each host uses $F(N)$ to develop a probability of a knowledgeable agent visiting it in the subsequent t iterations. The actual frequency of knowledgeable agent visits, $\frac{|A|}{t}$, is also recorded.

4.2   Experimental Results

The results we present are from 30 runs of the simulation, 300000 iterations each, with 30 hosts, 15 agents, and 3 instances of the service s.

Accuracy of PageRank. Figure 2 illustrates equation (4)'s accuracy in predicting the frequency of agent visits. One can see that the prediction approximates the actual value very closely and is also strongly correlated. The average coefficient of correlation between these variables over the set of 30 runs is 0.68. The average bias for the predicted probability is 0.01.

Verification of Services' Distribution. The frequency distribution for the number of instances of service s agents saw was recorded. The Shapiro-Wilk normality test returns a value of 0.5532 for the experimental distribution (with an infinitesimally small P-value), meaning the experimental distribution does partially deviate from normality. Nonetheless, this result implies that one *cannot* say that the data are *not* normally distributed. The deviation from normality can be explained by the low probability of an agent seeing an instance of the service; the data are skewed more toward an F-distribution. However, as demonstrated by the accuracy of $F(N)$ in Figure 3, it is reasonable to assume this distribution is normal.

Accuracy of $F(N)$. Figure 3 illustrates equation (10)'s accuracy in predicting the frequency of knowledgeable agent visits. One can see that the prediction approximates the actual value very closely and is also correlated. The average coefficient of correlation between these variables over the set of 30 runs is 0.60. The average bias for the predicted probability is -0.03.

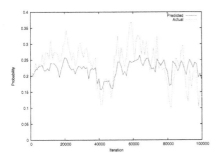
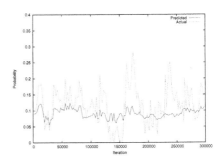

Fig. 2. Correlation between the predicted ν and actual agent frequency for the first 100000 iterations of simulation

Fig. 3. Prediction, $F(N)$, and actual agent visitation probabilities for a 300000 iteration simulation

5   Discussion

Examples of Applying the Technique. The information updates provided by the service discovery agents can be used to create new capabilities for multi-agent systems operating on peer-to-peer networks. Examples include:

- Learning availability thresholds, where if a service or host has not been "seen" by a discovery agent the remaining hosts and agents can (with high probability) assume that it has become unavailable and re-plan accordingly.
- Network triage, where the disappearance of discovery agents or their lack of contact with vital network nodes can be used to infer that the network has been damaged, compromised or segmented.
- Time-critical reasoning, where hosts use information provided by discovery agents to inform time-critical functions about how long they might reasonably expect to take to execute if they depend on remote services.
- Optimizing the number of agents, where, given an expected value for ν (which can be calculated using an expected network topology), one can use equation (10) to compute the optimal number of agents needed to achieve a required frequency of service discovery agent updates.

Limitations. Elements of the state description, N, can be both variable and unknown in some domains. Therefore agents might develop beliefs about the values of these parameters, such as ℓ and η. Furthermore, there are more efficient ways for the agents to traverse the network (i.e. self-avoiding walks). Research into these alternate techniques is required, however mathematically modeling them is more complex than with random walks.

Although ν can be defined from π_h, developing a belief of the global network topology, J, is a difficult problem on dynamic peer-to-peer networks. Pro-active routing algorithms for ad hoc networks often define a protocol to propagate this information, but this is expensive. The amount of memory/bandwidth required for each network message can in the worst case be $O(n^2)$ (to transmit the entire adjacency matrix). Since computation of $F(N)$ really only requires π_h (not the entire adjacency matrix J), π_h could be inferred by the observed frequency of agent visits at host h over a period of time. Research is required to evaluate this.

Future Work. We are currently implementing our approach for integration and testing in the Secure Wireless Agent Testbed (SWAT) [24]. This will include further empirical validation of the methods presented in this paper in a live testbed of at least a dozen mobile computing devices on an ad hoc wireless network.

Our proposed method for propagating service location information throughout the network can be used as a heuristic for mobile agent-based search. For example, hosts could cache data brought to them by the random walking service discovery agents. In doing so, each host would develop an index (or "belief") of the locations of services. These beliefs will become more accurate in conjunction with a host's proximity to the service. Therefore these beliefs, along with the timestamp of when the agent last saw the service, could be used as an A* search heuristic when locating the service. Work is needed to prove the feasibility, admissibility, and accuracy of this heuristic; we are in the process of using the SWAT to do so.

Further experimentation is required to ascertain the effect of varying parameters, such as ℓ. In addition, the effect of CPU and network bandwidth limitations is not clear. Finally, the accuracy of our approach on networks of heterogeneous hosts is unknown.

6   Conclusions

This paper addresses the critical problem of service discovery and location for multi-agent computing in dynamic, peer-to-peer networks. In this context, the location and capabilities of services, agents, and hosts are all part of a global state which can only be partially observed by each agent in the network. Our technical approach uses mobile agents and exploits the combinatorial properties of random walks to create a set of service discovery agents that maintain overall state for all nodes on the network. The principle contributions of this work include the development of a mathematical formulation of this problem of service discovery by mobile agents in a dynamic network and a set of empirical studies that validate the formulation.

Our results show that this pro-active approach can be used to maintain accurate state information across a dynamic network while having a limited effect on network messaging. We believe that this work represents an important example of how mobile agents can be practically adapted to the constraints posed by real network environments. In addition, this work can provide a basis for enabling multi-agent planning to sense and react to vital network-level events in order to improve plan execution and survivability.

References

1. Tambe, M.: Implementing agent teams in dynamic multi-agent environments. Applied Artificial Intelligence **12** (1998) 189–210
2. Kirousis, L.M., Kranakis, E., Krizanc, D., Stamatiou, Y.C.: Locating information with uncertainty in fully interconnected networks. In: International Symposium on Distributed Computing. (2000) 283–296
3. Perkins, C., Royer, E.M.: Ad-hoc on-demand distance vector routing. In: IEEE Workshop on Mobile Computer Systems and Applications. (1999) 90–100
4. Jacquet, P., Muhlethaler, P., Clausen, T., Laouiti, A., Qayyum, A., Viennot, L.: Optimized link state routing protocol for ad hoc networks. In: IEEE INMIC 01. Technology for the 21st Century. (2001) 62–68
5. Perkins, C., Bhagwat, P.: Highly dynamic destination-sequenced distance-vector routing (DSDV) for mobile computers. In: ACM SIGCOMM'94 Conference on Communications Architectures, Protocols and Applications. (1994) 234–244
6. Johnson, D.B., Maltz, D.A., Broch, J.: DSR: The dynamic source routing protocol for multihop wireless ad hoc networks. Ad Hoc Networking (2001) 139–172
7. Marwaha, S., Khong Tham, C., Srinivasan, D.: Mobile agents based routing protocol for mobile ad hoc networks. In: Proceedings of IEEE International Conference on Networks (ICON). (2002) 27–30
8. Minar, N., Kramer, K.H., Maes, P.: 12. In: Cooperating Mobile Agents for Dynamic Network Routing. Springer-Verlag (1999) ISBN: 3-540-65578-6.
9. Roy Choudhury, R., Bandyopadhyay, S., Paul, K.: A distributed mechanism for topology discovery in ad hoc wireless networks using mobile agents. In: Proceedings of the 1st ACM international symposium on Mobile ad hoc networking & computing, IEEE Press (2000) 145–146
10. Migas, N., Buchanan, W.J., McArtney, K.A.: Mobile agents for routing, topology discovery, and automatic network reconfiguration in ad-hoc networks. In: Proceedings of the 10th IEEE Conference and Workshop on the Engineering of Computer-Based Systems. (2003) 200–206
11. Langley, B., Paolucci, M., Sycara, K.: Discovery of infrastructure in multi-agent systems. In: Agents 2001 Workshop on Infrastructure for Agents, MAS, and Scalable MAS. (2001)
12. Kahn, M.L., Cicalese, C.D.T.: The CoABS grid. In: Innovative Concepts of Agent-Based Systems: 1st International Workshop on Radical Agent Concepts (WRAC). (2002)
13. Kozat, U.C., Tassiulas, L.: Service discovery in mobile ad hoc networks: An overall perspective on architectural choices and network layer support issues. Ad Hoc Networks **2** (2004) 23–44
14. Ratnasamy, S., Francis, P., Handley, M., Karp, R., Shenker, S.: A scalable content addressable network. In: Proceedings of ACM SIGCOMM 2001. (2001)

15. Stoica, I., Morris, R., Karger, D., Kaashoek, M.F., Balakrishnan, H.: Chord: A scalable peer-to-peer lookup service for internet applications. In: Proceedings of the 2001 conference on applications, technologies, architectures, and protocols for computer communications, ACM Press (2001) 149–160

16. Barrière, L., Flocchini, P., Fraigniaud, P., Santoro, N.: Capture of an intruder by mobile agents. In: Proceedings of the fourteenth annual ACM symposium on Parallel algorithms and architectures, ACM Press (2002) 200–209

17. Dasgupta, P.: Improving peer-to-peer resource discovery using mobile agent based referrals. In: Proceedings of the 2nd International Autonomous Agents and Multi-agent Systems Conference (AAMAS), Proceedings of the 2nd Workshop on Agent Enabled P2P Computing. (2003) 41–54

18. Nagi, K., Elghandour, I., König-Ries, B.: Mobile agents for locating documents in ad-hoc networks. In Gianluca Moro, C.S., Singh, M.P., eds.: Agents and Peer-to-Peer Computing (AP2PC 2003), Second International Workshop, Melbourne, Australia, July, 2003, Revised and Invited Papers. Volume 2872 of Lecture Notes in Computer Science., Springer (2004) 199–205

19. Peysakhov, M., Artz, D., Regli, W., Sultanik, E.: Network awareness for agent security in mobile ad-hoc networks. In: Proceedings of the Third International Joint Conference on Autonomous Agents and Multi Agent Systems, Association for Computing Machinery (2004) 368–375

20. Gkantsidis, C., Mihail, M., Saberi, A.: Random walks in peer-to-peer networks. In: Proceedings of the 23rd Annual Joint Conference of the IEEE Computer and Communications Societies INFOCOM '04 (to appear), IEEE (2004)

21. Brin, S., Page, L.: The anatomy of a large-scale hypertextual Web search engine. Computer Networks and ISDN Systems **30** (1998) 107–117

22. Sultanik, E.A., Peysakhov, M.D., Regli, W.C.: Agent transport simulation for dynamic peer-to-peer networks. Technical Report DU-CS-04-02, Drexel University (2004)

23. Barrett, C.L., Marathe, M.V., Engelhart, D.C., Sivasubramaniam, A.: Approximate connectivity graph generation in mobile ad hoc radio networks. In: Proceedings of the 36th Annual Simulation Symposium, IEEE Computer Society (2003) 81

24. Sultanik, E., Artz, D., Anderson, G., Kam, M., Regli, W., Peysakhov, M., Sevy, J., Belov, N., Morizio, N., Mroczkowski, A.: Secure mobile agents on ad hoc wireless networks. In: The Fifteenth Innovative Applications of Artificial Intelligence Conference, American Association for Artificial Intelligence (2003)

An Agent Module for a System on Mobile Devices

Praveen Madiraju, Sushil K. Prasad,
Rajshekhar Sunderraman, and Erdogan Dogdu

Department of Computer Science,
Georgia State University,
Atlanta, GA 30302
{cscpnmx, sprasad, raj, edogdu}@cs.gsu.edu

Abstract. A Middleware is the software that assists an application to interact or communicate with other applications, networks, hardware, and/or operating systems. We have earlier proposed an RMI-based middleware for mobile devices called System on Mobile Devices (SyD). A middleware on mobile devices is a challenging issue, as it has to deal with problems such as limited memory, frequent disconnections, low bandwidth connection, and limited battery life. The mobile agent module fits in the context of the middleware for mobile devices as it quite naturally alleviates the above mentioned problems. Communication between devices and method invocation capabilities, among other things are carried out by employing agents. In this paper, we provide the design and implementation of an agent module for SyD. We also present practical experiences gathered from carrying out experiments on the agent module.

Keywords: Agent Middleware for Mobile Devices, Agent Based Execution Engine, Mobile Agents, System on Mobile Devices Middleware.

1 Introduction

It has been widely acknowledged that a middleware is essential for application development on mobile devices. However, application on mobile devices introduces multiple challenges in a mobile setting. Mobile devices suffer from: frequent disconnection, low bandwidth connection, limited battery life, and limited memory. A middleware for mobile devices is a software that assists an application to interact or communicate with other applications on other mobile devices. A middleware should support the following basic set of services:

- *Communication Services:* enables communication between different mobile devices.
- *Execution and Listening Services:* provides capability to execute method calls on remote devices and also be able to listen to incoming method calls.
- *Data Access and Connectivity Services:* makes data on one mobile device be accessible to authorized groups of devices and also provide ability for mobile devices to connect to other devices.

G. Moro, S. Bergamaschi, and K. Aberer (Eds.): AP2PC 2004, LNAI 3601, pp. 144–152, 2005.

We have earlier proposed the System on Mobile Devices middleware (SyD) [11], [12], [13], [14]. SyD provides all the above mentioned services based on Remote Method Invocation (RMI). An emerging middleware approach is the agent oriented middleware approach. The services that a middleware provides, can be realized by employing mobile agents. Mobile agent approach inherently has advantages when compared to RMI. Once the agent is transported to a destination host, the agent can go ahead and execute even in case of a disconnection and when the connection is alive, the agent returns the result to the source host. This is the basic motivation for implementing an agent module for SyD.

In this paper we show the design and implementation of an agent module for SyD. One of the challenges of designing the agent module for mobile devices is the memory size overhead of the mobile agent framework. We have used μCode agent framework [10], as it is lightweight and also provides the basic required services of an agent framework. We also describe the experiments conducted on the agent module for SyD.

The rest of the paper is organized as follows: In Section 2, we give a brief overview of SyD middleware. We present background on agents and design of the mobile agent module for SyD in Section 3. The details of the experiments conducted on the agent module are described in Section 4. In Section 5, we compare our work with other peer's work and finally we offer our conclusions in Section 6.

2   SyD Middleware

We give a very brief overview of System on Mobile Devices (SyD) middleware [11], [12], [13], [14]. SyD is a new middleware technology that addresses the key problems of heterogeneity of device, data format and network, and that of mobility. SyD allows rapid development of a range of portable and reliable applications. SyD combines ease of application development, mobility of code, application, data and users, independence from network and geographical location, and the scalability required of large enterprise applications concurrently with the small footprint required by handheld devices. Each device is managed by a SyD deviceware that encapsulates it to present a uniform and persistent object view of the device data and methods. Groups of SyD devices are managed by the SyD groupware that brokers all inter device activities, and presents a uniform world view of the SyD application to be developed and executed on. All objects hosted by each device are published with the SyD groupware directory service that enables SyD applications to dynamically form groups of objects hosted by devices, and operate on them in a manner independent of devices, data, and underlying networks. The SyD groupware hosts the application and other middleware objects, and provides a powerful set of services for directory and group management, and for performing group communication and other functionalities across multiple devices. SyD middleware can function with or without a backbone network infrastructure on weakly connected networks as well as on ad-hoc networks providing varying levels of QoS guarantees.

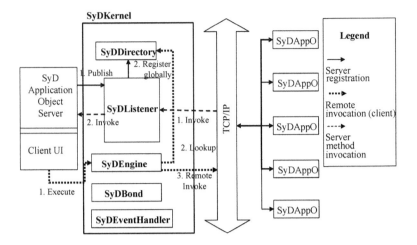

Fig. 1. SyD kernel architecture

We have earlier designed and implemented a modular SyD Kernel utility in Java. Fig. 1 describes SyD Kernel with the following five modules :

- **SyDDirectory:** Provides user/group/service publishing, management, and lookup services to SyD users and device objects and also supports intelligent proxy maintenance for users/devices.
- **SyDEngine:** Allows users to execute single or group services remotely and aggregate results.
- **SyDListener:** Enables SyD device objects to publish their services (server functionalities) as "listeners" locally on the device and globally via the directory services. It then allows users on SyD network to invoke single or group services via remote invocations seamlessly (location transparency).
- **SyDEventHandler:** Handles local and global event registration, monitoring, and triggering.
- **SyDBond:** Enables an application to create and enforce interdependencies, constraints and automatic updates among groups of SyD entities.

3   Agent Module for SyD

Here, we first give intoduction to mobile agents and then present the design of agent module in the context of SyD.

3.1   Mobile Agents

Mobile agents can be considered as an incremental evolution of the earlier idea of process migration. A mobile agent is an autonomous, active program that can move both data and functionality (code) to multiple places within a distributed system. The state of the running program is saved and transported to the new host, allowing the program to continue execution from where it left off before migration [5],[8].

Mobile agents require two components for their successful execution. The first component is the agent itself. The second component being the place where in an agent can execute. This is often referred to as the software agent framework. It provides services and primitives that help in the use, implementation and execution of systems deploying mobile agents. This generic framework allows the developers to focus on the logic of the application being implemented, instead of focusing on the implementation details of the mobile agent system. Specifically, It should support the creation, activation, deactivation and management of agents, which include mechanisms to help in the migration, communication, persistence, failure recovery, management, creation and finalization of agents. Additional services as naming and object persistence can also be provided. This environment must also be safe, in order to protect the resources of the machine from malicious attacks and possible bugs in the implementation of the agent code. Some of the popular examples are: IBMs Aglets[6], Mitsubishi Electric ITAs Concordia [7] and Object Spaces Voyager [4].

3.2   Design of Agent Module

In Fig. 2, we describe the agent module in the context of SyD. A mobile device serving as a server (SyD Application Object Server in Fig. 2), registers it's services and then provides services to clients. We illustrate this using the following steps (arrow labels with legend "Server registration"):

1. In the event of a new publish, the SyD Application Object Server sends a publish request to the Agent Module.
2. The Agent Module publishes and registers the services offered by the SyD Application Object Server to the SyDDirectory.

A mobile device serving as a client (Client UI in Fig. 2) can execute object services located on remote devices using the Agent Module. We illustrate the Client UI process in the following three steps (arrow labels with legend "Remote invocation(client)"):

1. The Client sends an execute of a remote service as a local call to the Agent Module.
2. The Agent Module dispatches an agent on to the SyDDirectory to get the remote user/service information of the remote server
3. With the user/service information (typically the URL of the remote server), the agent module dispatches another agent to complete the remote invocation.

We have used μCode[10] as our agent framework for mobile devices. Fig. 3 gives the internal details of the Agent Module. μCodeServer is running on each device listening for incoming mobile agents and is also capable of executing mobile agents on remote devices. Fig. 3 shows a sample method being executed by mobile device *1* on device *n*.

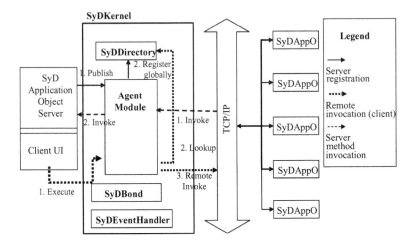

Fig. 2. Design of agent module for SyD

1. Mobile Device *1* sends an agent to the directory service to get the physical location information(URL) of device *n*.
2. The μCode agent framework has a listener, which listens for incoming mobile agents. Mobile Device *1* now has the physical location(URL) of the Mobile Device *n*.
3. Mobile Device *1* dispatches mobile agent to Mobile Device *n* to execute a method call.
4. Mobile Device *n* returns the result of executing the method call through an agent on to mobile device *1*.

However, it should be noted that, steps 1 and 2 from above could also be realized by simply exchanging messages between the agents rather than sending the agent by itself.

μCodeServer
μCode[10] is a lightweight software agent framework for mobile devices (small footprint - the core package is less than 18 Kbytes of jar file). It is a small Java API that aims at providing a minimal set of primitives to support mobility of code and state (i.e., Java classes and objects). It provides good abstractions for doing only a single thing, that is, moving code and state around. It also constitutes the kernel, providing small and efficient mechanisms for code mobility.

4   Experiments on the Agent Module

Experimental Test Bed: We ran our experiments on a high performance/low power SA-1110 (206 MHz) Compaq iPAQs H 3600, with 32 MB of SD RAM and 32MB of flash ROM. The handheld devices are connected through a mobile wireless network using a 2.4GHz wireless router. The operating system is

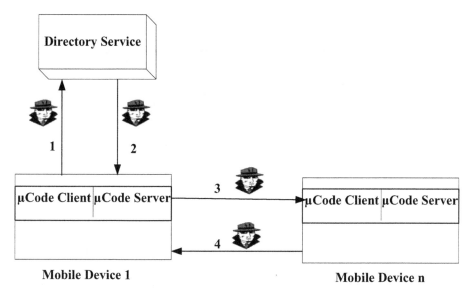

Fig. 3. Internal architecture of the agent module

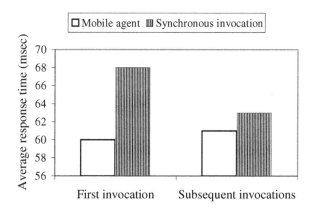

Fig. 4. Mobile agent vs Synchronous invocation - number of invocations

Windows CE. The mobile agent frame work on the iPAQ is μCode version 1.03. We use jdk version 1.3 to code our programs and JVM for iPAQ is Jeode VM Version 1.9. The DBMS of the directory server is Oracle 8i.

The proposed agent module for SyD replaces the SyDEngine-SyDListener pair. Here we compare synchronous invocation (SI) via SyDEngine-SyDListner pair with mobile agent (MA) using: response time based on multiple method invocations and response time based on size of data processed.

Response time is the total time required to execute a method call on a remote host. Number of method invocations is the number of times a particular method is called. In order to be consistent, we transferred a message of size 16 kilo bytes and the results of it are shown in Fig. 4. For a single method invocation, the

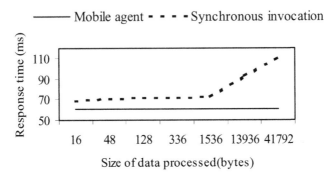

Fig. 5. Mobile agent vs Synchronous invocation - size of data processed

average response time of synchronous invocation (SI) of SyDEngine-SyDListner pair is much higher than mobile agent (MA) approach. However for subsequent method invocations, SI response time gets better. SI starts out with a higher response time and this could be attributed to the fact that, it needs to load stubs and skeletons and also factors such as marshalling/un marshalling and the implicit object serialization that is involved with any RMI-based approach. However for subsequent method calls, the client side stubs and skeletons are already bound to the server and RMI registry look up is faster. In the MA approach, the only initialization time is the time required for the μCode to start up. For all other subsequent method invocations, it's the same.

Fig. 5 gives the comparison based on the size of the data processed. The response time for MA is much lesser than SI. In the MA approach, data is processed at individual sites and processed data is sent across the network. In the SI approach, data is collected from multiple sites and then processing takes place on the gathered data and therefore results in higher response time.

5   Related Work

Our design of the mobile agent module is chiefly inspired from Limone[1]. Limone provides rapid application development over ad hoc network's consisting of logically mobile agents and physically mobile hosts. Lists of agents that satisfy the policy of host agent are stored into its acquaintance list. The host agent retains full control of the local tuple space since all remote operations are simply requests to perform a particular operation for a remote agent and are subject to policies specified by the operation manager. This high degree of security encourages a collaborative type of interaction among agents. This coordination model and middleware promises to reduce development time for mobile applications. We don't have the concept of a tuple space in our model description. However limone uses tuple space, as it's underlying model. Our directory service serves the purpose of the acquaintance list.

A programmable event based middleware[3] is developed for pervasive mobile agent organizations. A concept of organization oriented framework for the design of mobile agent application in pervasive computing scenarios is discussed. This middleware is an event-based approach based on the definition of a minimal event-kernel, which is suitable for deployment in resource-constrained devices.

A mobile agent based PC Grid[2] is a mobile agent based middleware where remote computers users wish to mutually offer their desktop computing resource to other internet group members. Each agent represents a client user, which carries out their requests, searches for the available resources, executes the job at suitable computers, and migrates it to others when the current ones are not available.

Data Lockers[15] is another research activity under mobile agent middleware. Data lockers allow users of mobile devices to rent space at the fixed network. This helps mobile users to perform computations remotely with out bothering about the memory space and computation capacity of mobile devices.

A plenitude of research is available on mobile agent based approach for middleware as discussed in the programmable event based middleware[3], PC Grid[2] and Data Lockers[15]. However, we have not seen much research comparing the different middleware approaches. A close line of study to ours can be found in [9]. They discuss performance evaluations of different java based approaches to web database access. We compare middleware approaches and [9] compares java based approaches.

6   Conclusions

We already have a full scale design and implementation of a RMI-based middleware (SyD). We proposed the design and implementation of an agent module for SyD. We have implemented, evaluated and compared the agent module versus the synchronous invocation of SyDEngine-SyDListener Pair. We have also presented performance comparisons of average response time based on varying number of method invocations. We have not taken in to account of the security drawbacks that mobile agents imposes on the system. The security aspect is ignored in this paper as it is out of the scope for the performance evaluations.

As part of the future work, we plan to carry out experiments and do performance evaluations based on: response time(n), where n is the no. of disconnections and the agent framework overhead vs the SyDListener overhead. We aim to design and implement a hybrid engine that extracts the best of the features of agent and RMI approaches by automatically switching between them depending on a decision algorithm.

References

1. C.-L. Fok, G.-C. Roman, and G. Hackmann. A lightweight coordination middleware for mobile computing. Technical report, Technical Report WUCS-03-67, Washington University, Department of Computer Science and Engineering, 2003.
2. M. Fukuda, and Suzuki N. Tanaka, Y., L.F. Bic, and S. Kobayashi. A mobile-agent-based pc grid autonomic computing. In *Fifth Annual International Workshop on Active Middleware Services (AMS'03)*, pages 696 – 703, Seattle, Washington, June 25 - 25, 2003.

3. M. Gazzotti, M. Mamei, and F. Zambonelli. A programmable event-based middleware for pervasive mobile agent organizations. In *11th IEEE EUROMICRO Conference on Parallel, Distributed, and Network Processing*, pages 517–525, Genova, Feb. 2003.

4. G. Glass. Overview of voyager: Objectspace's product family for state-of-the-art distributed computing. Technical report, ObjectSpace, 1999.

5. C. G. Harrison, D.M. Chessm, and A. kershenbaum. Mobile agents: Are they a good idea? Technical report, Research Report, IBM Research Division, 1994.

6. G. Karjoth, D. Lange, and M. Oshima. A security model for aglets. *IEEE Internet Computing*, 1(4), 1997.

7. R. Koblick. Concordia. In *Communications of the ACM*, march, 1999.

8. P. Madiraju and R. Sunderraman. A mobile agent approach for global database constraint checking. In *ACM Symposium on Applied Computing (SAC'04)*, pages 679–683, Nicosia, Cyprus, 2004.

9. S. Papastavrou, P.K. Chrysanthis, G. Samaras, and E. Pitoura. An evaluation of the java-based approaches to web database access. *International Journal of Cooperative Information Systems*, 10(4), 2001.

10. Gian Pietro Picco. μcode: A lightweight and flexible mobile code toolkit. In *Mobile Agents, Procs. of the 2nd Intl. Workshop on Mobile Agents (MA)*, volume 1477, pages 160–171. Springer, LNCS, Stuggart, 1998.

11. Sushil K. Prasad, V. Madisetti, et al. System on mobile devices (SyD): Kernel design and implementation. In *First Intl. Conf. on Mobile Systems, Applications, and Services (MobiSys), Poster and Demo Presentation*, San Francisco, May 5-8, 2003.

12. Sushil K. Prasad, V. Madisetti, et al. Syd: A middleware testbed for collaborative applications over small heterogeneous devices and data stores. In *5th ACM/IFIP/USENIX International Middleware Conference*, Toronto, Ontario, Canada, October 18th - 22nd, 2004.

13. Sushil K. Prasad, Vijay Madisetti, et al. A middleware for collaborative applications over a system of mobile devices (SyD): An implementation case study. Technical report, Technical Report CS-TR-03-01, Department of Computer Science, Georgia State University, July 16, 2003. `http://www.cs.gsu.edu/~cscskp/PAPERS/CONF/TechRep/SyDTechReport.doc`.

14. Sushil K. Prasad, M. Weeks, et al. Toward an easy programming environment for implementing mobile applications: A fleet application case study using SyD middleware. In *IEEE Intl Workshop on Web Based Systems and Applications, at 27th Annual Intl. Computational Software and Applications Conf. (COMPSAC)*, pages 696 – 703, Dallas, Nov 3-6, 2003.

15. Y. Villate, A. Illarramendi, and E. Pitoura. Data lockers: Mobile-agent based middleware for the security and availability of roaming users data. In *IFCIS International Conference on Cooperative Information Systems (CoopIS'2000)*, September, 2000.

Multi-agent System Technology for P2P Applications on Small Portable Devices

Martin Purvis, Noel Garside, Stephen Cranefield,
Mariusz Nowostawski, and Marcos De Oliveira

Department of Information Science, University of Otago,
Dunedin, New Zealand
(mpurvis, ngarside, scranefield, mnowostawski,
moliveira)@infoscience.otago.ac.nz

Abstract. In this paper we discuss architectural design issues and trade-offs in connection with our experiences porting our agent-based platform, Opal, to the Sharp Zaurus personal digital assistant (PDA). At the present time, the Zaurus is able to run the Java-based Opal platform with RMI, HTTP and JXTA (but not JXME) as message transports. There were many adjustments that had to be made in order to establish JXTA functionality over Java Personal Profile on the Zaurus systems, but it may be an easier process in the future if some of these changes are incorporated into the JXTA standard. The wireless and Bluetooth capability of the Zaurus make it ideal for bridging the gap between Bluetooth networks and traditional networks. The extension of mobility to distributed Agent-based systems will be a significant growth area in future agent research, and the Zaurus PDA a glimpse into the future functionality that mobile distributed agent applications may provide. We also discuss how Opal's unique support for micro agents may facilitate the deployment of advanced agent systems on future medium- and small-footprint devices.

Keywords: MAS technology, portable devices, JXTA, P2P.

1 Introduction

Developments in and widespread deployment of telecommunications and distributed system technology have led to increased interest in the ideas of multi-agent systems [1]. Since the notion of software agents represents an embodiment of a distributed and autonomous form of peer-to-peer (P2P) computing, the practical deployment of multi-agent systems will be facilitated if the agent system technology can interoperate with standard infrastructural P2P services wherever possible. We have made progress in this area by developing Opal [2], a standard agent-based software platform in Java that provides support for the agent communication protocols specified by the Foundation for Intelligent Physical Agents (FIPA) [3], and extending the Opal platform [4] so that it can be used in conjunction with JXTA [5], a standard for P2P interactions.

G. Moro, S. Bergamaschi, and K. Aberer (Eds.): AP2PC 2004, LNAI 3601, pp. 153–160, 2005.

In this paper we discuss a further issue associated with agents and peer-to-peer computing: the use of this technology in the context of small hand-held devices employing wireless communications. In particular, we describe the design trade-offs associated with the deployment of multi-agent system and P2P technology (in our case, with the Opal+JXTA system) onto personal digital assistants (PDAs). In this connection, consideration must be given to a number of down-to-earth issues in order to realize a practical agent-based P2P system on a PDA physical platform.

2   Opal and JXTA

The Opal FIPA Platform includes the KEA micro-agent framework [4,6]. At a high level intelligent software agents can be treated as individual FIPA-compliant (they employ the FIPA ACL communication protocols) agents. Individual tasks within such agents are delegated to appropriate micro-agents. This approach offers the advantage of reusing components, together with late dynamic binding between particular roles.

2.1   JXTA

Opal has been built to conform to the latest specification of the FIPA Abstract Architecture (FIPA AA). At the present time the Transport Service, as specified in the FIPA AA, is used solely to provide a communication protocol for ACL messages between two end-points. But the Transport Service does not cover some aspects of agent communication, such as discovery, multicasts or broadcasts. Since these were needed for our application, we implemented them using our own proprietary interfaces and protocols. To facilitate the dynamic discovery of peers on the network and peer-to-peer messaging, we use the JXTA infrastructure [5], which is a set of open protocols that allow any connected device on the network to communicate and collaborate in a P2P manner. Thus the standard set of transport protocols in OPAL (IIOP and HTTP) has now been extended to include JXTA. In addition to the protocols, Project JXTA also maintains an up-to-date open-source Java implementation of the JXTA protocols. We use this JXTA source code, because it has been developed and validated by the Project JXTA community. In this section we describe how P2P JXTA communication can coexist with FIPA-prescribed agent communication.

A peer in JXTA represents any networked device that implements one or more of the JXTA protocols. To send messages, peers use pipes, which offer asynchronous, unidirectional message transfer for service communication. JXTA advertisements are metadata structures (in XML) used to describe and announce the existence of peer resources. Peers discover resources by searching for their corresponding advertisements and may cache any discovered advertisements locally.

The Project JXTA community has also been working on providing implementations of portions of the JXTA protocols for the Java 2 Mirco Edition (J2ME), and this work has been referred to as JXME [5]. We will discuss JXME specifics in context below.

2.2 Messaging

Messaging at the lowest micro-agent level is implemented using method calls, and its semantics is expressed simply by method calls signatures. At a higher level, micro-agents employ a limited communication model of communication, based on the notion of goals, declarations, and commitments, with the semantics expressed by UML models of goals and their relationships. At the highest level agents use standard FIPA ACL, augmented with the notion of object-oriented ontologies represented in UML [7].

Since the FIPA ACL does not currently have a notion of an agent group, and thus has no notion of a public announcement to a group, JXTA communication can fill the gap and play a useful role. We use a special service agent, called a Peer agent, to facilitate this process: there is a single Peer agent for each JXTA peer (i.e. a single Peer agent per machine). Communication between Peer agents is performed by means of JXTA announcements and pipes (thus outside normal FIPA ACL messaging). Practically, this means that public announcements are done via JXTA announcements and ordinary peer-to-peer communication is performed via standard FIPA messaging mechanisms transmitted via the JXTA Pipe infrastructure. All the public announcements are done in an asynchronous (and unreliable) manner over the standard JXTA Content Advertisements. Since the Peer also has a standard Pipe for FIPA text-based ACL messaging, all communication can be considered to be performed over JXTA.

In our P2P implementation there is an additional transport layer between the FIPA agent and the ordinary (FIPA-compliant) Transport System. This layer is provided by the specialist Peer agent, which intercepts messages from individual agents and propagates them appropriately for the P2P environment. For messages addressed to a single individual agent registered on the local peer, the Peer agent simply forwards the message directly to the recipient. If the receiver is registered on a remote peer, the local Peer agent passes the message to that recipient. Peer agent, which in turn passes the message down to the individual recipient. If, however, the original message is a public announcement, then the local Peer agent passes the announcement to all locally registered agents and also passes it to all other Peer agents, which in turn pass it down to all their local agents. The Peer agent is implemented on a level below the FIPA ACL level, so its communications are not entirely based on the FIPA ACL itself, but rather on a proprietary protocol implemented on the OPAL platform.

3   The PDA Platform

The PDA platform that we used is the Sharp Zaurus SL-c700 hand-held device. Standard versions include a VGA (640x480) colour screen, 64 MB of RAM, and can run the Java Personal Profile (Java PP) version of J2ME under the Linux operating system. Wireless communication with other devices can be performed via WiFi (IEEE 802.11b networking [8]) and Bluetooth [9].

The Sharp Zaurus represents a platform with power midway between that of current workstations and cell phones. Although it is more powerful than current cell phones, its memory resources and Java JDK are limited when compared with standard computers. However, its existing computing resources may provide a useful testing environment for cell phone on the horizon [10].

3.1 Zaurus Resources: JDK, Memory, Bandwidth, and Power Requirements

The Zaurus runs Personal Profile J2ME, which is the equivalent of JDK1.3 minus deprecated methods and Swing. While Swing can be run on the Zaurus, it runs very slowly and takes more memory than is desirable. Although the Zaurus has 64 MB of flash RAM for storing files and running programs, it appears to have only about 8 MB of RAM for running Java applications. The Zaurus also has expansion slots on its chassis for compact flash (CF) and Security Digital (SD) RAM cards. The CF slot is used for both the wireless and Bluetooth cards, so SD RAM was used for secondary storage. As a result, it is possible to run a 30 MB JAR file off of the SD RAM card.

Bandwidth under different networks varies widely, and thus the amount of agent messaging that can be performed in agent applications under different network configurations must be taken into account. For example when employing JXTA networking, a small hand-held device running JXME would have to poll larger devices (that are capable of running full JXTA) for relevant messages rather than receiving all the messages on the network and filtering them itself. This polling situation would require more message exchanges.

The use of WiFi networking is demanding for the electrical batteries available for the the Zaurus, and the Zaurus can run for little more than one hour when not on mains power. This means that in the near term Bluetooth networking, with its lower battery demands, may be a better wireless network option than 802.11.

4   Opal Agent Platform on the Sharp Zaurus

Our porting approach was to attempt, as a proof of concept, to port an agent application that was built with Opal (an electronic trading application discussed in [4]) that included Opal and the developed application, which employed a Petri net tool [11], onto the Zaurus. The agent application was intended to feature dynamic (runtime) compilation of Petri nets, so the code for the custom Java dynamic compiler was altered in an to attempt to allow dynamic compiling on the Zaurus.

4.1   JXTA and Java Personal Profile (Java PP)

Since JDK 1.4 incorporates more security features (than JDK 1.3), it may be difficult to keep JXTA compatible with Java PP. One suggested approach might be to have support for JDK1.3 a core feature in JXTA, with additional security features left to the individual developers. Since JXTA is open source, it would be possible to re-engineer the source files and change the security code when required, but this option would become difficult to maintain as new JXTA functionality is added. Though it may be inappropriate to restrict full JXTA to JDK1.3 without deprecated methods and Swing, it may be feasible to create a middle level between JXTA and JXME, which we might call JXPP. The primary advantage of a JXPP would be the recognition of a middle tier that accommodates the rising capabilities of newer mobile telephones and PDAs, which need an intermediate step between JXTA and JXME. JXPP would have all the current functionality of JXTA, but would not limit the development of JXTA in terms of using JDK1.4 or evenJDK1.5. It could be developed by people interested in rising efficiency technologies rather than leading-edge full power technology. It could possibly represent those who are concerned with producing the current technology, but in more compact packages and lower costs.

4.2 Opal Supporting Agent Advertisements to Groups

JXTA supports sending messages to groups, whereas FIPA does not specify sending messages to groups of agents. Since Opal is modular (transport is a separate modular layer), each transport used should also support sending messages to groups. Under RMI, groups can be simulated by opening registries listening on different ports, e.g. 1099, 1100, etc. Under HTTP subnets can simulate groups, with multicasting used to send messages to all members of each subnet. P2P networks groups are significant, since multicasting is an efficient and lightweight message transport, but they may not be scalable if the number of peers on the subnet are not kept to within reasonable limits. The extension of the FIPA ACL protocols to include groups is an ongoing area of research [4].

5 Micro-agents and MIDP

One approach for reducing agent application size for operation on smaller devices is the use of micro-agents [6]. Micro-agents are a finer-grained implementation of agents, which communicate using method calls rather than declarative representations of FIPA-specified speech acts that then employ lower-level transports, such as IIOP. Micro-agents can run in one thread, rather than having separate threads for each the agent object, which could potentially lead to a significant overhead savings on for applications running on devices operating under the MIDP [12] version of J2ME (typical cell phones, for example).

5.1 Issues Constraining the Minimal Size for a Multi-agent Platform on PDAs

To achieve a flexible architecture for multiple platforms, the intention was to establish core classes for J2ME, with supplemental classes for J2SE. This would involve removing the Swing-based graphical user interface from the core classes, removing references to JDK1.4 from core classes, and removing calls to JDK1.3

deprecated methods. The underlying transports JXTA, IIOP, RMI, and HTTP all add size to the OPAL platform, but are more elegant, flexible and robust than listening on ports. By reducing the ported application to packages that only the Zaurus uses, the JAR file was reduced to 7 Mb. Reducing Opal all the way down to MIDP presents some interesting challenges in terms of reducing Opal's size. JXTA and HTTP will work as transports, but RMI and IIOP will not, since they require too many resources. Since JXTA now operates on TINI [13] devices, it should work on MIDP mobile phones. Bluetooth automatically discovers other devices within range, so peer discovery should not be problematic. Bluetooth scatternets are made up of up to ten personal-area networks (PANs), or piconets. A piconet has a master and up to seven slaves, where all communication is through the master, and slaves never communicate directly. The Zaurus would make a good wireless P2P group master, administering the smaller slave devices and then communicating via wireless to an infrastructural network. Bluetooth devices, such as TINI devices and mobile telephones, will probably only poll for services like JXTA.

One change that would facilitate the porting of Opal to MIDP concerns the XML parser in Opal. Opal currently uses a fully featured XML parser, Xalan [14], which is 1.6 MB in size. Changing to a J2ME XML parser, such as MinML [15] (14 KB in size) would be appropriate for smaller devices, such as those running Java MIDP. Although these CML packages may not have all the features required for Opal, J2ME-specific code could be developed to use them on the Zaurus or other small devices. This is a change we intend to make in the near future.

6   Discussion

Current PDAs, e.g. the Zaurus C7x0, can run a full JXTA (or Agent) platform, which means that the periphery of the network has now been pushed to mobile telephones, and embedded devices (using, for example, TINI [13]). Mobile phones and other MIDP devices have limited storage, memory and bandwidth (especially limited by their slow front side busses) but can still run JXME (e.g. JXME works on TINI devices). A possible P2P micro-device architecture could involve a MIDlet able to run on a mobile phone that could act as a container or mini platform for KEA micro-agents. The MIDlet could communicate with a JXTA relay running on a Zaurus device. This would allow an implementation with application-specific code that removed overheads (such as XML parsers) e.g. parts of agent messages could be passed to the MIDP device as strings. It would also make it easier to provide a graphical user interface to create agent messages.

6.1   Mobile P2P Agent Application Architectures

The mobility of software agents, as envisioned in the customary sense, involves an agent moving physically from one device to another by using a transport, such as RMI. There have not been many applications, however, for which this physical movement has been identified as required or significant, because an agent could simply communicate with another agent almost anywhere directly via, say, HTTP, rather than having to move from one device to the next. In connection with PDAs, like the Sharp Zaurus, there is a sense of mobility associated with the ability of the Zaurus to move to different places. This is significant for Bluetooth and wireless networks: the Zaurus can move to within a remote Bluetooth PAN (or Piconet) and then communicate with agents there, which may not have access to a transport like HTTP. Fixed Bluetooth devices also have a geographical location, so it can be inferred that the Zaurus is within 10 -100 metres of the Bluetooth devices within range. This type of mobility offers many advantages for a PDA like the Zaurus: it can visit and administer remote Piconets. If a piece of machinery has a TINI device fitted to it, it may be able to communicate exceptions to the Zaurus that will allow the problem to be quickly fixed or reset. In workflow situations, jobs could be assigned to the closest workers by their position relative to a bluetooth location. Location awareness could also trigger changes to the an application interface running on a PDA, e.g. an inventory application could be automatically enabled in the warehouse, while a customer service application is activated on the shop floor.

The Zaurus in a JXTA-connected agent application could act as a bridge between a Bluetooth network and a fully featured wireless network. Since the 802.11b bandwidth is not as high as some broadband networks, the agents on the Zaurus PDA could poll agents on the broadband network for relaying to agents on the Bluetooth network, rather than them receiving all the messages on the network. The Bluetooth agents could in turn poll the Zaurus for their specific messages, thus reducing the flow of messages to a reasonable amount in terms of the local network type.

7   Conclusion

The Sharp Zaurus proved fully capable of running the entire full-sized Opal platform, and the major transports used for agent messaging, JXTA, HTTP, and RMI, were fully operational. This means that P2P agent-based computing can incorporate hand-held wireless devices like the Zaurus, and it was not necessary to create an entirely separate Opal release branch for running on Zaurus-sized devices. We intend to restructure (repartition) the Opal internal architecture in order to use J2ME components for loading the platform onto the Zaurus-type devices, and this should facilitate more sophisticated agent-based P2P applications to be run in distributed wireless environments.

Our work indicates that now mid-range (in terms of power) devices, such as Sharp Zaurus PDAs, can operate as P2P rendezvous nodes, while the smaller MIDP-running devices can operate as individual peers. The Opal agent system currently allows for full-sized agents on the Zaurus devices and smaller (possibly only KEA micro-agents) on the MIDP devices. Thus using the notion of JXTA P2P architectures and hierarchically refinable agents, agent architectures can be effectively spread across a wide range of wireless computing architectures.

References

1. Jennings, N.R.: Agent-oriented software engineering. In: Proceedings of the 12th International Conference on Industrial and Engineering Applications of AI. (1999)
2. Purvis, M., Cranefield, S., Nowostawski, M., Carter, D.: Opal: A Multi-Level Infrastructure for Agent-Oriented Software Development. Technical Report 2002/01, University of Otago, Dunedin, New Zealand (2002) ISSN 1172-6024.
3. Foundation For Intelligent Physical Agents (FIPA): Fipa 2001 specifications. http://www.fipa.org/specifications/ (2003)
4. Purvis, M., Nowostawski, M., Cranefield, S., Oliveira, M.: Multi-Agent Interaction Technology for Peer-to-Peer Computing in Electronic Trading Environments. In Moro, G., Sartori, C., Singh, M., eds.: Second International Workshop on Agents and Peer-to-Peer Computing, Second International Joint Conference on Autonomous Agents and Multiagent Systems (AAMAS 2003), Melbourne Australia (2003) 103–114
5. Project JXTA: Jxta specifications. http://www.jxta.org (2003)
6. Nowostawski, M., Purvis, M., Cranefield, S.: KEA - Multi-level Agent Infrastructure. In: Proceedings of the 2nd International Workshop of Central and Eastern Europe on Multi-Agent Systems (CEEMAS 2001), University of Mining and Metallurgy, Krakow, Poland (2001) 355–362 http://www.sf.net/projects/javaprs.
7. Cranefield, S., Purvis, M.: A UML Profile and Mapping for the Generation of Ontology-specific Content Languages. Knowledge Engineering Review, Special Issue on Ontologies in Agent Systems (2002) 21–39
8. http://grouper.ieee.org/groups/802/11/ (2004)
9. http://www.bluetooth.com/ (2004)
10. http://www.infoworld.com/article/03/02/13/HNmotolinux 1.html/ (2003)
11. Nowostawski, M.: Jfern, version 1.2.1. http://sf.net/projects/jfern (2002)
12. http://www.micronova.com/ZAURUS/index.html/ (2004)
13. http://www.ibutton.com/TINI/, http://tini.jxta.org// (2004)
14. http://xml.apache.org/xalan-j// (2004)
15. http://www.wilson.co.uk/xml/minml.htm/ (2004)

Coordinator Election Using the Object Model in P2P Networks

Hirokazu Yoshinaga, Takeshi Tsuchiya, and Keiichi Koyanagi

Graduate School of Information, Production, and Systems, Waseda University,
2-9, Hibikino, Kitakiyushiyu, Fukuoka, Japan
yoshinaga@akane.waseda.jp, tuchiya@suou.waseda.jp,
keiichi.koyanagi@waseda.jp

Abstract. We propose the ACE (Adaptive Coordinator Election) platform that elects and relocates a coordinator adaptively in P2P networks.
In collaborative applications, a coordinator mediates synchronization,
consistency, sequencing and delay difference. However, it is difficult to
decide a coordinator in applications used in P2P networks because of
some characteristics of network, e.g., network instability, and differences
in physical networks and devices for participants. The ACE platform
elects and relocates a coordinator dynamically according to environmental metrics obtained from participants. Each metric has a priority and
weight to allow a coordinator to be chosen according to the purpose of
applications. We implemented our platform using the JXTA framework
and tested it. The results show the feasibility of adaptive coordinator
relocation in P2P networks.

1 Introduction

The Internet environment has dramatically changed beyond our expectations
and is progressing toward resource ubiquity such as network connections, computing devices, and contents in edge devices. A ubiquitous network in which all
devices can connect to networks and get information at anytime is expected. In
that network, devices in different physical networks will interconnect with each
other and can share information or collaborate. Some applications can satisfy
some of those requirements, but not as much as we would like. Groove [2] enables users to share information such as schedules or files using a group space.
Users can connect with each other via a fixed server. Gnutella[3] enables users to
share files by transferring messages among users without central servers. It leads
to an active use of information in an environment with ubiquitous resources.
Collaborative tools like MSN Messenger [4] and AOL Messenger [5] enable users
to communicate in real time via a central server. Nowadays people increasingly
want to share information without being dependent on their physical networks
in such a ubiquitous resource environment. We have developed an application
for collaboration called JXCube (Jxta eXtreme Cube) in the JXTA [6] project.
This is a fully distributed collaborative application and it enables users to work

G. Moro, S. Bergamaschi, and K. Aberer (Eds.): AP2PC 2004, LNAI 3601, pp. 161–172, 2005.

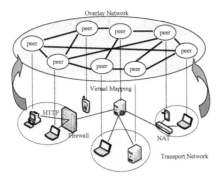

Fig. 1. Overlay Network

together without a fixed central server. JXCube [7] offers a secured collaborative work space through the use of user groups and encrypted messages, plug-in collaborative functions, and replication of users work space having same identity. This paper shows the architecture for relocating the node that mediates collaborative work (coordinator) in fully distributed environments such as P2P networks.

2 Collaborative Work in a P2P Network

2.1 P2P Networks

We stand for a peer-to-peer(P2P) network [1] as a distributed networking technology in the application layer of the TCP/IP reference model. It is possible to construct an overlay network using UUID (Universally Unique Identifier) instead of using IP addresses on a physical network (Fig. 1). That is, P2P offers a logical network on top of underlying networks. Even if users are in different networks and use different transport protocols, they can communicate with each other via the overlay network. For example, if one user can use only HTTP in a company and another can use TCP at home, they can communicate with each other via a logical communication path. In that scenario, a node with a global IP address that offers HTTP service act as a broker for these two, enabling them to communicate with each other. Also, users use not only a wired network with fixed PCs, but also a wireless network like 802.11x with mobile devices such as laptop PCs, PDAs, and cell phones. These mobile devices fit into a P2P network by using overlay functions, even if their topology changes continually.

JXTA and Gnutella support some kinds of overlay functions. JXTA is a set of P2P protocols and a framework. It is consisted of edge peers, relay peers, and rendezvous peers. Edge peers are the most common. A relay peer forwards messages on behalf of a peer that cannot directly address another peer (e.g., in NAT/firewall environments), bridging between different physical and/or logical networks. A rendezvous peer maintains resource information (advertisements) that an edge peer requests to find other peers, groups, and communication pipes.

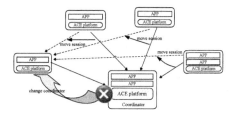

Fig. 2. Dynamic Relocation of an Adaptive Coordinator

And also, JXTA supports peer group that is a collection of peers that have agreed upon a common set of services. On the other hand, Gnutella is a P2P protocol that was basically developed for file sharing. It constructs a self-organized network of peers transferring messages among themselves (flooding). A new version of Gnutella (Gnutella2) now has scalability for routing by supporting two types of peers: leaf nodes and hub nodes. Leaf nodes are edge nodes forming the network. Hub nodes maintain resources for leaf nodes to communicate with other leaf nodes. The characteristics of a P2P network are as follows.

(1) The network is constructed by multi-hop routing.
(2) The network is unstable because changes in dynamic topology caused by nodes joining or leaving.
(3) All nodes act equally. There are no explicit roles.
(4) It covers a heterogeneous transport environment where device types, capabilities, and communication methods are different.

2.2   Collaborative Work on a P2P Network

It is necessary to mediate work for synchronization, consistency, sequencing, and delay differences in applications such as schedule management, videoconferencing, online games, and other collaborative applications. For examples, a schedule management application must assure consistency among users. Also, a videoconferencing application must mediate delay differences to keep communication consistency. That is, it is necessary for collaborative applications to transfer some events such as working events or result events in the same sequence or at the same time. In this paper, we define these issues as a mediation problem. In previous collaborative work on a network, there were two types of methods for resolving it: the server model and the distributed model. The server model collects messages to a one node from users although it may be overloaded. On the other hand, in the distributed model, each node must propagate event messages to all nodes by broadcasting or multicasting. This increases of network traffic. And each node must execute complicated mediation work. We chose the server model because we assume that collaborative worker number only up to a few dozen people so the load will be low; it thus is easy to maintain event messages and it is simple to implement it.

3   Adaptive Dynamic Relocation of Coordinator

3.1   Characteristics of ACE Platform

The basic model for ACE (Adaptive Coordinator Election) platform that we propose is shown in Fig. 2. The basic requirements are follows.

(1) Optimal coordinator selection for collaborative applications
(2) Transparent relocation of a coordinator for collaborative applications
(3) Crash coordinator detection automatically
(4) Low dependence on OS and collaborative applications

It is necessary to decide the optimal coordinator according to environmental information that changes dynamically such as routing information by user participation, secession, and user movement, network use rate, CPU and memory use rate for each device, battery remaining rate and so on (requirement 1). Also, it is necessary for users to keep using applications so they do not have to reconnect when the coordinator is replaced (requirement 2). And it is necessary to keep using applications without stopping the system when a coordinator crashes (requirement 3). Moreover, it is possible to correlate with various collaborative applications by separating this platform from them (requirement 4). The relocation of a coordinator treated in the ACE platform can be regarded as a one of the leader election problem in the distributed system of old models such as [8], [9], and [10]. For this problem, an algorithm elects the only node; a lot of research has been done on this. Moreover, some research [11] and [12] treats the leader election problem on an ad hoc network. Reference [11] describes an algorithm that elects a leader using multicast and reference [12] shows an algorithm in which the node located at the center of the topology becomes a leader in an ad hoc network. Those studies do not meet the requirements mentioned above. ACE platform is built to satisfy the point of electing the leader (coordinator) in P2P networks including a wide area network based on environmental information about nodes that composes a network.

3.2   Basic Operation

ACE platform operates according to the following procedure when a user newly participates in or leaves from a collaborative work space (Fig. 3).

(1) Each node searches for a current coordinator using ALM (Application Level Multicast).
(2) A coordinator responds to the request if it exists.
(3) Each node acquires environmental information at constant intervals.
(4) Each node transmits environmental information to a coordinator.
(5) The coordinator sends an ACK message to each node.
(6) The coordinator elects a new coordinator and a candidate one.
(7) The old coordinator transmits the relocation data to the new one.
(8) The new coordinator notifies all members of the change.

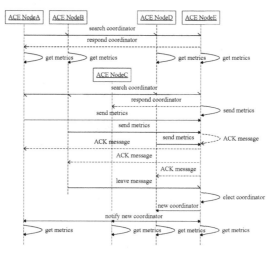

Fig. 3. Basic sequence of ACE platform

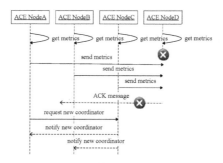

Fig. 4. Crash detection of coordinator

In (1), the node that transmitted a request message first becomes the initial coordinator because there is no response to the request when the user first participates in the collaborative work space. At this time, it is necessary to set the waiting time for responses to request messages beforehand. A coordinator waits until all nodes have transmitted environmental information in (4). At this time, if environmental information is not transmitted from a participated node even after the fixed waiting time has passed, it elects a new coordinator considering that went off-line because of a crash. The waiting time is set dynamically using the RTT (round trip time) values from an participants.

3.3 Crash Detection and a Coordinator Relocation Technique

The ACE platform offers a mechanism for detecting a coordinator crash without stopping the system. This section describes the operating procedure when a coordinator crashes (Fig. 4).

(1) One node does not send an ACK message after a fixed time has passed for an environment metric transmission.
(2) That node is considered to have node crashed, so the node that discovered it asks for a candidate coordinator.
(3) The new coordinator notifies all nodes of the change.

The waiting time of the ACK message is set dynamically using the RTT values of participants. Also, candidate coordinators are selected according in ascending order in (2). After (3), the process continues to basic operation (3) in the previous paragraph.

3.4   Discussion About Environmental Information

The ACE platform dynamically decides a coordinator among users according to the network and device status. This section discusses environmental information (metrics) considered in ACE platform. It is possible to divide it roughly into network-dependent metrics and device-dependent metrics as dynamically changing information. The ACE platform uses the following metrics.

– Topology location
– Network usage rate
– CPU usage rate
– Memory usage rate
– Battery remaining rate
– Continuous network connection time

Network dependent metrics are topology location and network usage rate. The topology location is a metric that becomes effective when users move frequently. When users move frequently and the network topology changes dynamically, it is possible to have a uniform number of hops and a response time to a coordinator among nodes if central node becomes it. When users move, routing information changes although user identity (overlay address, logical address) does not change because of the overlay function. It is possible to distinguish user mobility by getting routing information about each node. In the ACE platform, topology location is calculated according to routing information and a coordinator then decides a central node. The metric of network usage rate is effective when a physical network where a coordinator exists is overloaded because a message delay will be generated between the coordinator and each node. In such a situation, coordinator relocation can shorten the response time. In the ACE platform, the average RTT is used to measure network usage rate. Device dependent metrics include CPU usage rate, memory usage rate, and a battery remaining rate. CPU and memory usage rates are effective when the load on the coordinator is high, because mediation work and response time will worsen. In such a situation, coordinator relocation can make the system better. The ACE platform uses these average values to avoid temporary loads. The battery remaining rate is effective when users are outside. For such a situation, it is necessary to reduce battery

```
# Ace Mediation Policy
Metric.Period = 50000

# Priority of Metric
Metric.Num = 6
Metric.Priority1 = ace.metrics.Location
Metric.Priority2 = ace.metrics.RTT
Metric.Priority3 = ace.metrics.BatteryPower
Metric.Priority4 = ace.metrics.MemoryUsage
Metric.Priority5 = ace.metrics.CPUUsage
Metric.Priority6 = ace.metrics.ContinuousTime

# Coefficient of priority
Coefficient.Priority1 = 0.9
Coefficient.Priority2 = 0.7
Coefficient.Priority3 = 0.5
Coefficient.Priority4 = 0.4
Coefficient.Priority5 = 0.3
Coefficient.Priority6 = 0.1
```

Fig. 5. Mediation policy definition file

consumption so the user can work for a long time. In particulars, when participants are only mobile devices, it is better for a device with a high battery level to become a coordinator. In the ACE platform, mobile devices will be targets. It is necessary to consider the network connection times of nodes as an environmental metric. If a device that has not worked for a long time becomes a coordinator, traffic increases when the coordinator is relocated and then many messages are transmitted in every time. It is possible to solve this problem by having a node that has been connected for a long time becomes a coordinator. Metric dependencies will be different for each application because there are a lot of environmental metrics for nodes in a P2P network. The ACE platform can decide the best node by considering several metrics with priority levels.

3.5 Mediation Policy Definition Technique

There are various kinds of collaborative applications such as a videoconferencing, schedule management, and online games, and the purpose of using a coordinator is different for each application. For example, videoconferencing or online games use a coordinator to mediate delays and sequences among nodes. In that case, if a node connected with a physical network where throughput is low becomes a coordinator, it will cause bottlenecks. When all nodes are using mobile devices, it is also necessary to consider the topology location and the battery level. Moreover, a schedule management application needs a lot of processing power to mediate consistency. In this case, system performance will improve if a node with low CPU or memory use rate becomes a coordinator. That is, it is necessary to define a policy for deciding a coordinator because the purpose of using a coordinator is different for different applications. The ACE platform resolves this problem by considering a few metrics. It defines the number of metrics, the priority level, and the weight coefficient for each metric as shown in Fig. 5. The number of metrics determines the number of environment metrics, the metric priority levels set the priority of metrics corresponding to the purpose of the coordinator, and the weight coefficients set the weight put on the priority of metrics.

3.6   Coordinator Election Technique

In this section, we discuss the technique for electing a new coordinator. First, the current coordinator decides the node order for each metric using those values it received from all nodes. The following procedure shows how to decide the central node order of the topology.

(1) The coordinator makes a connected graph using routing information transmitted by all nodes. When relay nodes that are not participants exist, the number of these nodes is allocated as the edge weight.
(2) The coordinator makes a minimum spanning tree using a width priority search that makes itself the starting point and excludes closed paths.
(3) All leaf nodes are removed from that minimum spanning tree. Also, all leaf nodes are removed from the partial tree. This is repeated until the partial tree has only one node.

The node order is in descending order from the center, and closed path (2) or leaf nodes (3) become new orders in order of those appearing. The network, CPU and memory usage rates are sorted in ascending order using values from each node. Also the battery level and continuous network connection time are sorted in descending order. Next, the coordinator and candidate new coordinators are elected from the ordered node by following procedure.

(1) Order weight W_k is added to the node order for each metric

$$W_k = \frac{N+1-k}{N} \qquad (k : order, N : nodenumber) \qquad (1)$$

(2) Weight coefficient C_{mj} is applied to each metric and metric score $S_{i,j}$ is calculated for each node.

$$S_{i,j} = W_k \times C_{mj} \qquad (i : node, j : metric) \qquad (2)$$

(3) The sum of metrics score SUM_i is calculated for each node.

$$SUM_i = \sum^{j} S_{i,j} \qquad (3)$$

(4) The node with the highest sum is elected as the new coordinator NEW and the others become candidate coordinators.

$$NEW = \max \{SUM\} \qquad (4)$$

If the elected coordinator is different from a current one, the change is notified to all members. Also, candidate coordinators are used when the current coordinator would crash.

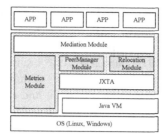

Fig. 6. ACE Service software architecture (APP:application)

4   Implementation

4.1   ACE Service

The ACE platform is implemented by dividing it from various collaborative applications. As a result, it can handle various applications without implementing complicated mediation logic in each application. Figure. 6 shows software architecture of the ACE platform. The ACE platform was written in the Java language to make it applicable to various OSs. It uses the JXTA framework to compose a P2P network on a Java virtual machine and is implemented as a JXTA peer group service called ACE Service. Peer group service is composed of a collection of instances of the service running on multiple members of the peer group. If any one peer fails, the collective peer group service is not affected. We implement it as a peer group service for all nodes to use. ACE Service is composed of four modules: Relocation Module, Peer Manager Module, Metrics Module, and Mediation Module. The adaptive dynamic coordinator relocation proposed in this paper is composed of only Relocation Module, Peer Manager Module and Metrics Module without Mediation Module. The Mediation Module offers mediation functions for i) synchronization, ii) consistency, iii) sequencing, or iiii) delay differences.

4.2   Modules

The Relocation Module provides functions for electing a coordinator and relocating it. Functions are implemented according to Basic operation of section 3.2, Crash detection and a coordinator relocation technique of section3.3, and Mediation policy definition technique of section3.5 and Coordinator election technique of section3.6. The Peer Manager Module is used only by a coordinator, and it manages the presence and status of participants as a list. A coordinator adds or updates the status of nodes online when a node joins a peer group and sends its message. It also updates their status offline when a node leaves a peer group and sends its message. Moreover, a coordinator removes their status from the list, if it cannot receive their metric values after waiting for a predetermined time. The Metrics Module provides a series of functions for operating environment metrics and ranking them. It is composed of some objects that each stand for a metric. And these objects are implemented as distributed objects [13] that contain data and a series of procedures for operating it. Separating metrics from other modules enables us to extend metrics and change their implementations easily. All environment metrics use the same interface (Metric), so that these metrics can be operated easily. Also, abstract definitions of metric classify each environment metric according to that feature and simplify metric implementations. For example, the topology location metric and the network use rate metric are defined as network metrics, and they defines a each communication method. Metric objects are transmitted to a coordinator after the metrics values for each node have been obtained. The coordinator ranks these objects with a ranking procedure.

4.3 Handling the Delay Between Nodes

If participants are distributed widely in a physical network, then nodes far from a coordinator experience delays in sending or receiving messages compared with nearby nodes. Also, if nodes are in an environment where only http communication is accepted, there is a delay caused by the pooling time or protocol exchange on a relay node. Thus, there are delay differences between the coordinator and nodes in communication in an overlay network. Therefore, a coordinator may judge that a node with a delay cannot send a message and the node joining a collaboration work space. Even if a long waiting time is set, it is difficult to decide that value because we cannot predict which network nodes are present. In ACE Service, the coordinator's waiting time is the maximum time (response time of the most delayed node) of RTT from each node plus a fixed time. This time is also used as the waiting time when each node receives an ACK message from the coordinator.

5 Evaluation

We evaluated our model by testing whether it could elect a coordinator adaptively. In this experiment, we use the CPU metric and the Memory metric as environmental metrics. The intervals for getting metric values for each node were set to 50 s and the weight coefficients were set to 0.9 for CPU metric and 0.6 for Memory metric. Figure. 1 shows experimental environment we used. And then all node are connected with wired LAN using 100BaseTX. Figure. 7 stands for the transition of RTT values between a coordinator when adaptive dynamic relocation was used (Pattern1) and not used (Pattern2). The message length of RTT was set to be 1 KB. In this experiment, three nodes (ACE NodeA, ACE NodeB, and ACE NodeC) participated in a collaborative work space. ACE NodeA became the coordinator in the first election in both patterns. In Pattern1, the average RTT value from the beginning of the experiment until the load applied was 371 ms, the value after 100 s of applying load to NodeA was 1206 ms, and the value after relocation was 490 ms. That is, when the load was applied, the RTT value from NodeC to the coordinator (NodeA) was about 81% higher than

Table 1. Experimental machine

	ACE Node A	ACE Node B	ACE Node C
CPU	Pentium4 2.6GHz	PentiumM 1.4GHz	Pentium4 800MHz
RAM	1124MB	512MB	512MB
OS	Windows 2000 Professional	Windows XP Professional	Windows XP Professional ! !
JDK	Java2Platform SE v.1.4	Java2Platform SE v.1.4	Java2Platform SE v.1.4

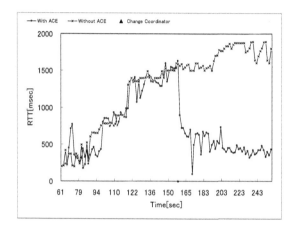

Fig. 7. Progress of RTT value by relocation

before the load was applied to the coordinator. And NodeB which has a low CPU usage rate, because the new coordinator after the next election. The RTT value from NodeC to the coordinator (NodeB) decreased by about 61% compared with after the load was applied. However, the memory usage rate hardly changed for each node.

On the other hand, in Pattern2, the average RTT value from the beginning of the experiment to before load application was 413 ms and the value 100 s after the load was applied to NodeA was 1487 ms. That is, when load was applied, the RTT value from NodeC to the coordinator (NodeA) increased by about 90% compared with before the load was applied to a coordinator. As a result, the average RTT value was shortened to 997 ms using the adaptive dynamic relocation compared with not using it. Although it took 60 s for the coordinator to be relocated after applying the load, it is possible to relocate the coordinator in the early stages during an overload by decreasing the metric acquisition and election intervals.

6   Conclusion

In this paper, we proposed a method of dynamically relocating of the coordinator which mediates synchronization, consistency, sequencing, and delay differences in a P2P network without an explicit server. Our model elects a coordinator using environmental metrics which change dynamically for each node. We implemented our model as ACE Service using the JXTA framework and evaluated it. The results confirmed that ACE Service elects an optimal coordinator among nodes and relocates it adaptively according to environmental information. In the near future, we will implement the Mediation Module and test ACE Service using collaborative applications.

Acknowledgements

This work is partly supported by a specific project of WASEDA university $2003A - 950$. Also part of this research is joint research with NTT DoCoMo.

References

1. Keiichi Koyanagi, Takashige Hoshiai, and Hidekazu Umeda, Proposal and Introduction on P2P Networking Technologies, IEICE Transactions on Communications, VOL.J85 $- B$, NO.3, pp.319 $- 332$, 2002 $- 3$.
2. Groove Network, inc.: Groove Product Backgrounder, Corporate whitepaper, 2002.
3. Gnutella: http://www.gnutella.com/
4. MSN Messenger : http://messenger.msn.co.jp/
5. AOL Messenger: http://www.jp.aol.com/aim/
6. Project JXTA: http://www.jxta.org/
7. Project JXCube: http://jxcube.jxta.org/
8. Hector Garcia-Molina: Elections in a distributed@computing system, IEEE Transactions on Computers, C-31(1) : 47 $- 59$, January 1982.
9. Singh G., Leader Election in the Presence of Link Failures: IEEE Trans, Parallel and Distributed Systems, Vol.7, No.3, pp.157 $- 171$, 1996.
10. Fetzer, C., and Cristian, F.: A Highly Available Local Leader Election Service, IEEE Trans Softw Eng, Vol.25, No.5, pp.603 $- 618$, 1999.
11. Royer, E.M., and Perkins, C.E.: Multicast Operations of the Ad-hoc On-Demand Distance Vector Routing Protocol, Proc 5th Annual ACM/IEEE International Conference on Mobile Computing and Networking (MOBICOM), pp.207 $- 218$, 1999.
12. Suzuki, Y., Ishihara, S., and Mizuno T.: Relocation of a Mediation Function on a Mobile Ad Hoc Network, IPSJ, Vol.43, No.12, pp.3959 $- 3969$, Dec. 2002.
13. Nakajima, T., Aizu, H., Kobayashi, M. and Shimamoto, K.: Environment Server: A System Support for Adaptive Distributed Applications, Lecture Notes in Computer Science, Vol.1368, pp.142 $- 157$, 1998.

The Dynamics of Peer-to-Peer Tasks: An Agent-Based Perspective

Xiaolong Jin[1], Jiming Liu[1], and Zhen Yang[2]

[1] Department of Computer Science, Hong Kong Baptist University,
Kowloon Tong, Hong Kong
{jxl, jiming}@comp.hkbu.edu.hk
[2] Electrical Engineering Section, The School of Railway Mechanism of Lanzhou,
Lanzhou, 730000, China
yieytmz2@hotmail.com

Abstract. Grid computing aims at integrating geographically distributed computers and providing 'super-supercomputers' that can be seamlessly accessed by users all over the world. In peer-to-peer grids, numerous tasks are distributed to grid nodes in a decentralized fashion. In this case, two issues of interest are suitable computing mechanisms and the global performance of the grid, specifically, the dynamics of task distribution and handling. To address these issues, in this paper we present an agent-based adaptive paradigm for peer-to-peer grids and further identify two typical scenarios corresponding to task distribution and handling in this paradigm. We provide two models to characterize the agent-based scenarios. Based on our characterizations, we identify the key features of, and the effects of, several important parameters on the dynamics of task distribution and handling in peer-to-peer grids.

1 Introduction

As a departure from traditional IT, grid computing aims at integrating and sharing distributed computer resources and thus providing 'super-supercomputers' for users all over the world [1,2,3]. Because grids usually integrate numerous distributed computers and sometimes it is impossible to organize them in a centralized architecture, grids in peer-to-peer architectures [1] attract a lot of attention from researchers. Peer-to-peer grids consist of numerous grid nodes, and the submitted tasks are distributed to grid nodes without a centralized control or scheduling mechanism. Thereby, two key issues of research on peer-to-peer grids are (1) finding a computing mechanism suitable to situations with large-scaled grid nodes and tasks and (2) examining the dynamics of peer-to-peer grids as well as emergent global behaviors. This paper will address these issues.

1.1 Peer-to-Peer Computing Architecture

Peer-to-peer computing has quickly emerged as a new paradigm for developing distributed, Web-based systems [4]. Peer-to-peer systems consist of distributed

G. Moro, S. Bergamaschi, and K. Aberer (Eds.): AP2PC 2004, LNAI 3601, pp. 173–184, 2005.

and decentralized components, called *nodes*, which usually have the same roles, responsibilities, and symmetric communications among them [5]. Such systems are mainly designed to share files among distributed computers at the beginning stage of peer-to-peer computing. In [4], Ge et al. designed a mathematical model for characterizing the behavior of peer-to-peer file sharing systems. Currently, researchers have begun to explore new applications of peer-to-peer computing. Many large projects, such as JXTA [6], have been started during the past several years. In addition, research on peer-to-peer grids [1] is also one of the important branches of peer-to-peer computing.

1.2   Agent-Based Peer-to-Peer Systems

Agent-based systems have been widely used in peer-to-peer computing [5,7]. It is regarded as a perfect match to integrate peer-to-peer computing and agent-based systems [5,8], because "since their inception, multi-agent systems have always been thought of as networks of peers" [8]. By observing the perfect match as well as the potential benefits, Li *et al.* have developed an agent-based platform, called *A-peer*, to facilitate the deployment of agents in a peer-to-peer environment [7]. In peer-to-peer grids, it is locally determined, according to some requirements, to distribute tasks to grid nodes. This can be implemented by agents and their local behaviors. Based on this observation, in this paper, we present an agent-based peer-to-peer computing paradigm in grids.

1.3   Problem Statements

As we mentioned above, in peer-to-peer grids, continuously submitted tasks are distributed to numerous grid nodes in a decentralized fashion. Facing such a situation with large-scaled tasks and grid nodes and hence involving some uncertainty, what we primarily concern are how to provide a scalable computing mechanism for peer-to-peer grids and then how to examine their global dynamics in depth. This paper will address these two issues. Specifically, the latter issue can be investigated in two typical scenarios: (1) Given a short time interval during which no new tasks are submitted and no old tasks are handled, can we provide a macroscopic model to characterize the dynamics of task distribution? (2) Given a long time interval during which some new tasks are submitted and some old tasks are handled at each step, can we provide a model to characterize the dynamics of task handling?

1.4   Organization of the Paper

The rest of the paper is organized as follows: In Section 2, we describe an agent-based grid computing paradigm. In Sections 3 and 4, we present two macroscopic models to characterize the dynamics of task distribution and handling in two identified scenarios. Through case studies, we validate the effectiveness of our models. We also show our observations and analyze the corresponding mechanisms. Section 5 concludes the paper and presents the directions for future work.

2 Agent-Based Computing Paradigm in Peer-to-Peer Grids

In this section, we will provide an agent-based computing paradigm in peer-to-peer grids. Our computing paradigm and following models are inspired by the differential equation based modeling work in [9,10].

In our paradigm, we employ agents to carry tasks. Then, the movements of agents correspond to the transports of tasks among grid nodes. When a task is submitted to a grid, an agent will be generated to carry it. Each agent carries a task and wanders on the network of grid nodes to search for an idle node to form a new agent team[1], or an existing agent team to join and queue. Here, we define the period of time that an agent spends on wandering before it joins an idle node or an existing agent team as time delay. An agent prefers to join a short team rather than a long team because of the possibly long waiting time. Specifically, in our models, we set a maximum size for agent teams. Agents will not join teams with the maximum size. Globally, how many wandering agents join teams of a certain size depends not only on the number of currently wandering agents, but also on the numbers of currently existing teams of various sizes. In this sense, our models are adaptive.

After having joined a team, an agent can also decide to leave the team and move to other nodes because, as we have mentioned above, it prefers to queue at a short team. Note that in the proposed paradigm, agents are assumed to be memoryless. In other words, the experience of an agent will not affect its following behaviors. According to the above description, an agent has three main behaviors: *wandering*, *queuing*, and *leaving*. In order to have its task handled, an agent must be served by one of the grid nodes. An agent only has local information about the sizes of teams, where it is queuing or which it encounters while wandering. It does not have the global knowledge of the whole grid. After an agent queues at the first place of a team for a unit of service time, its task will be handled and then the agent itself will disappear from the gird environment automatically.

It should be pointed out that this paper will focus on a peer-to-peer minigrid environment the same as in [11], where (1) grid nodes are homogeneous, and provide the same services; and (2) tasks are divided into independent subtasks with the same size before they are submitted to the grid. In addition, we assume that (1) agents follow the same strategies of wandering, queuing, and leaving; and (2) time delay and service time are positive constants.

3 The Dynamics of Task Distribution

This section will provide a macroscopic model to describe agent-based task distribution scenario, where the total number of tasks as well as the total number of agents remain unchanged over time. This means, during the process of task distribution, (1) no new tasks are submitted to the grid environment. Hence, no new agents are generated; (2) no tasks are finished. Accordingly, no agents disappear.

---

[1] In the paper, we refer to the queuing agents at a grid node as an agent team.

3.1   Characterizing the Process of Task Distribution

Our following model will focus on characterizing the number of wandering agents as well as the number of agent teams of different sizes. Let q_0 be the number of wandering agents, q_s be the number of agent teams of size s, and m be the maximum team size. Then, we should have $q_s \geq 0$ $(0 \leq s \leq m)$. This is a prerequisite of our model. According to the above description of task distribution scenario, we have the following general model:

$$\frac{dq_1(t)}{dt} = j_0 q_0(t) - j_1 q_1(t) + l_2 q_2(t) - l_1 q_1(t),$$

$$\frac{dq_s(t)}{dt} = j_{s-1} q_{s-1}(t) - j_s q_s(t) + l_{s+1} q_{s+1}(t) - l_s q_s(t) \quad (1 < s < m), \quad (1)$$

$$\frac{dq_m(t)}{dt} = j_{m-1} q_{m-1}(t) - l_m q_m(t),$$

$$\frac{dq_0(t)}{dt} = \sum_{s=1}^{m} l_s q_s(t) - \sum_{s=0}^{m-1} j_s q_s(t),$$

where coefficients j_s $(0 \leq s < m)$ and l_s $(0 < s \leq m)$ are adaptive with the real-time agent distribution. They are functions as follows:

$$0 \leq j_s \Big(q_0(t), q_0(t - \tau); q_1(t), \cdots, q_{m-1}(t) \Big) \leq 1 \text{ and } 0 \leq l_s \Big(q_s(t) \Big) \leq 1. \quad (2)$$

In (1), the first three equations characterize the quantitative changes of teams of size 1, s $(1 < s < m)$, and m, respectively; The last equation characterizes the quantitative change of wandering agents. To better understand the equation system, here we will give some more detailed descriptions. First, let us see the second equation, which is a general one. In the second equation,

1. $j_{s-1} q_{s-1}(t) - j_s q_s(t)$ describes the quantitative change of teams of size s caused by wandering agents' joining at certain teams. Specifically, $j_{s-1} q_{s-1}$ (t) denotes that there are $j_{s-1} q_{s-1}(t)$ teams of size $s - 1$, each of which has an agent beginning to wander from time $t - \tau$ [2,3] joining at time t. Then, these teams become teams of size s. Therefore, the number of teams of size s will increase with $j_{s-1} q_{s-1}(t)$. $j_s q_s(t)$ is similar.
2. $l_{s+1} q_{s+1}(t) - l_s q_s(t)$ describes the quantitative change of teams of size s caused by queuing agents' leaving. Specifically, $l_{s+1} q_{s+1}(t)$ denotes that at time t, the last agents at $l_{s+1} q_{s+1}(t)$ teams of size $s + 1$ leave. Accordingly, these teams become teams of size s. The number of teams of size s will increase with $l_{s+1} q_{s+1}(t)$. $l_s q_s(t)$ is similar.

The first equation is a special case, where $j_0 q_0(t)$ denotes that $j_0 q_0(t)$ wandering agents from time $t - \tau$ meet idle nodes at time t and form new teams of

---

[2] We refer to an agent beginning to wander from time $t - \tau$ as an wandering agent from time $t - \tau$.

[3] Here, time $t - \tau$ is manifested in the coefficient function $j_{s-1}(\cdot)$.

size one. Then, the number of teams of size one increases with $j_0 q_0(t)$. As compared with the second equation, the third equation misses two terms, $-j_m q_m t$ and $l_{m+1} q_{m+1}(t)$, because there are no teams of size $m + 1$ and no agents will join teams of size m. The fourth equation describes the quantitative change of wandering agents where, $\sum_{s=1}^{m} l_s q_s(t)$ denotes the total number of agents that leave teams of various sizes; $\sum_{s=0}^{m-1} j_s q_s(t)$ denotes the total number of wandering agents from time $t - \tau$, which meet certain existing teams and join them to queue at time t, or meet idle nodes and form new teams of size one at time t.

Now, let us explain the meanings of $j_s(\cdot)$ and $l_s(\cdot)$ $(0 \le s \le m)$ in detail:

1. $j_s(q_0(t), q_0(t - \tau); q_1(t), \cdots, q_{m-1}(t))$ denotes the rate of teams of size s, which agents, beginning to wander at time $t - \tau$, join at time t. The reason why $j_s(\cdot)$ is related to $q_0(t)$ and $q_0(t - \tau)$ as well as $q_1(t), \cdots, q_s(t), \cdots, q_{m-1}$ (t) is that at time t, some wandering agents from time $t - \tau$ meet idle nodes and form teams of size one, or join existing teams of various sizes. General speaking, the more teams of size s, the more chances with which wandering agents join them, thus the larger j_s. Therefore, rate j_ss should be determined by the numbers of wandering agents at time $t - \tau$ and t, as well as the numbers of various teams. Here, $\tau > 0$ denotes time delay.
2. $l_s(q_s(t))$ denotes the rate of teams of size s, whose last agents leave at time t. It only depends on the number of teams of size s at time t.

According to their specific meanings, $j_s(\cdot)$ and $l_s(\cdot)$ $(0 \le s \le m)$ are subject to the following constraints:

1. Constraints about $\{j_s\}$ $(0 \le s \le m)$:

$$\sum_{s=0}^{m-1} j_s \cdot q_s(t) \le q_0(t) \quad \text{and} \quad \sum_{s=0}^{m-1} j_s \cdot q_s(t) \le q_0(t - \tau). \tag{3}$$

The above two constraints denote that the number of wandering agents from time $t - \tau$, which form new teams of size one or join existing teams of various sizes at time t, should be less than or equal to the total number of wandering agents from time $t - \tau$ as well as the number of wandering agents at time t.

2. A constraint about j_s and l_s $(0 \le s \le m)$:

$$0 \le j_s \cdot q_s(t) + l_s \cdot q_s(t) \le q_s(t). \tag{4}$$

where $j_s \cdot q_s(t)$ denotes the numbers of teams of size s, which have wandering agents joining; $l_s \cdot q_s(t)$ denotes the numbers of teams of size s, where the last agents leave. The above constraint indicates there is only a part of teams of size s, where either some wandering agents join or the last agents leave.

3.2 Case Studies

In this subsection, we will through case studies validate that our model is effective in characterizing the process of task distribution. Meanwhile, we will examine the dynamics of task distribution.

For the sake of illustration, this subsection will set $m = 2$. According to the constraints on j_ss and l_ss discussed in the previous subsection, we can set parameter functions $j_0(t)$, $j_1(t)$, $l_1(t)$, and $l_2(t)$ as follows:

$$j_1(t) = \begin{cases} 0, & \text{if } f(t,\tau) = 0 \text{ or } q_1(t) = 0, \\ p_1 \cdot f(t,\tau)/q_1(t), & \text{if } q_1(t) > f(t,\tau), \\ p_2, & \text{otherwise,} \end{cases} \tag{5}$$

$$j_0(t) = \begin{cases} 0, & \text{if } f(t,\tau) = 0, \\ p_4 \cdot (1 - j_1(t)q_1(t))/f(t,\tau), & \text{otherwise,} \end{cases} \tag{6}$$

$$l_1(t) = p_3 \cdot (1 - j_1(t)), \text{ and } l_2(t) = p_5. \tag{7}$$

where $f(t,\tau) = min(q_0(t - \tau), q_0(t))$, p_i $(i = 1, \cdots, 5)$ can be functions with a range of $[0,1]$ or constants in $[0,1]$. Without loss of generality, in the following case studies, we set them as constants.

Non-negativeness and Global Stability. Since $q_0(t)$ and $q_s(t)(1 \leq s \leq m)$ denotes the number of wandering agents and the number of teams of size s, respectively, they should be non-negative. Moreover, according to [12], the process of load balancing will tend to a steady state finally. In the following, we will through a case study show that our task distribution model possesses the above two properties, namely,

1. Non-negativeness: $q_0(t), q_1(t)$, and $q_2(t)$ remain non-negative as $t > 0$.
2. Global Stability: $q_0(t), q_1(t)$, and $q_2(t)$ tend to unique steady states as t increases.

Case Study 1. *In this case study, we set $S(0) = 1000$, $\tau = 0$, $p_1 = 0.1$, $p_2 = 0.1$, $p_3 = 0.1$, $p_4 = 0.3$, and $p_5 = 0.2$. It contains five cases with different initial agent distributions: Case 1: $q_0(0) = 1000$, $q_1(0) = 0$, and $q_2(0) = 0$; Case 2: $q_0(0) = 0$, $q_1(0) = 1000$, and $q_2(0) = 0$; Case 3: $q_0(0) = 0$, $q_1(0) = 0$, and $q_2(0) = 500$; Case 4: $q_0(0) = 100$, $q_1(0) = 100$, and $q_2(0) = 400$; Case 5: $q_0(0) = 300$, $q_1(0) = 300$, and $q_2(0) = 200$.*

From the results shown in Figure 1, we can note that:

1. For all cases, their $q_0(t)$, $q_1(t)$, and $q_2(t)$ curves remain non-negative during the process of task distribution. For each case, its $q_0(t), q_1(t)$, and $q_2(t)$ curves monotonously converge to steady states. All cases take around 30 time units to converge to a unique steady state where $q_0(t) = 210$, $q_1(t) = 580$, and $q_2(t) = 105$. According to the setting of p_i, we can note that averagely $j_0(t)$ is greater than $j_1(t), l_1(t)$, and $l_2(t)$. That means agents prefer to form teams of size one, rather than wander or queue at teams of size two. This is the reason why at the final steady state, there are relatively more agent teams of size one.

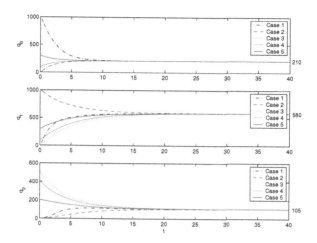

Fig. 1. Case Study 1: Task distribution

2. Let us take Case 4 as an example to show the working mechanism of our model. At the beginning, as compared with the numbers of wandering agents and agent teams of size one, there are too many agents queuing at teams of size two. Therefore, some queuing agents at those teams leave and become wandering agents. At the same time, those teams become new teams of size one. Here, we should note that while some queuing agents at teams of size two leave, there are also some wandering agents that join existing teams of size one and form new teams of size two. However, the number of the former is relatively greater than that of the latter. This explains why $q_2(t)$ curve decreases whereas $q_0(t)$ and $q_1(t)$ curves increase gradually.

Remark 1. Through the above and other case studies (not presented), we have the following observations: (1) No matter what the initial agent distributions are, the numbers of wandering agents and queuing agents at various teams always keep non-negative during the process of task distribution, and finally converge to unique steady states; (2) Given a setting of parameters p_1, p_2, p_3, p_4, and p_5, all different initial agent distributions always converge to a unique balanced and steady state; (3) The final steady distribution depends on the setting of p_1, p_2, p_3, p_4, and p_5. Different settings lead to different final agent distributions; (4) The number of steps taken to converge depends on the setting of p_1, p_2, p_3, p_4, and p_5.

Time Delay. In the following, we will examine the effects of time delay on the process of task distribution.

Case Study 2. *In this case study, we set* $S(0) = 1000$, $p_1 = 0.1$, $p_2 = 0.1$, $p_3 = 0.1$, $p_4 = 0.3$, $p_5 = 0.2$, $q_0(0) = 100$, $q_1(0) = 100$, *and* $q_2(0) = 400$.

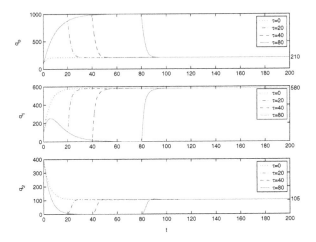

Fig. 2. Case Study 2: Time delay

The results are shown in Figure 2. For the sake of space limitation, we will not elaborate on the details of the figure.

Remark 2. In light of Case Study 2, we can observe the following phenomena: (1) Even if taking into account time delay, the final agent distribution, i.e., the final steady state of task distribution, does not change with different values of time delay; (2) Time delay plays a linear role. In the case of considering time delay, the time units that the process of task distribution takes to converge are the summation of time delay and the time units needed in the case of not considering time delay.

4   The Dynamics of Task Handling

In this section, we try to characterize the dynamics of task handling in the task handling scenario, where (1) there are some new tasks submitted to the grid environment at the beginning some steps. Accordingly, the same number of agents are generated to the grid environment; (2) at each step, some old tasks, which have queued at the first places of certain teams for a unit of service time, are finished. Therefore, the corresponding agents disappear from the grid environment automatically. As compared with the task distribution scenario, the task handling scenario is more general and more realistic. The task distribution scenario can be regarded as a relatively small locality of the task handling scenario.

4.1   Characterizing the Process of Task Handling

Given the above description, we can extend our task distribution model to the following new one:

$$\frac{dq_1(t)}{dt} = \dot{j}_0 q_0(t) - \dot{j}_1 q_1(t) + l_2 q_2(t) - l_1 q_1(t) + f_2 q_2(t) - f_1 q_1(t),$$

$$\frac{dq_s(t)}{dt} = \dot{j}_{s-1} q_{s-1}(t) - \dot{j}_s q_s(t) + l_{s+1} q_{s+1}(t) - l_s q_s(t) + f_{s+1} q_{s+1}(t) - f_s q_s(t),$$

$$\frac{dq_m(t)}{dt} = \dot{j}_{m-1} q_{m-1}(t) - l_m q_m(t) - f_m q_m(t), \tag{8}$$

$$\frac{dq_0(t)}{dt} = \sum_{s=1}^{m} l_s q_s(t) - \sum_{s=0}^{m-1} \dot{j}_s q_s(t) + \begin{cases} g(t) & t \leq T \\ 0 & t > T \end{cases},$$

where,

1. $f_{s+1} q_{s+1}(t) - f_s q_s(t)$, called "task handling" term, describes the quantitative change of teams of size s, because some queuing agents at the first places have been handled. Specifically, $f_{s+1} q_{s+1}(t)$ denotes that $f_{s+1} q_{s+1}(t)$ teams of size $s+1$ at time $t - \lambda$ become teams of size s, because the tasks carried by their first agents are handled. Therefore, the number of teams of size s will increase with number $f_{s+1} q_{s+1}(t)$. Here, $\lambda > 0$, called *service time*, denotes the time units to finish a task.

2. $T \geq 0$, called *time threshold*, denotes the time point before which new tasks are submitted to the grid and accordingly new agents are generated to carry them.

3. $g(t)$ denotes the number of new tasks submitted at time $t < T$.

Specifically, in (8), f_s ($1 \leq s \leq m$) has the following form:

$$0 \leq f_s(q_s(t), q_s(t - \lambda)) \leq 1, \tag{9}$$

where denotes that the rate of agent teams of size s, which have a task finished, is dependent on the numbers of agent teams of size s at time $t - \lambda$ and t. In other words, $f_s(\cdot)$ is adaptive. According its meaning, f_s ($1 \leq s \leq m$) has the following constraints:

$$f_s \cdot q_s(t - \lambda) < q_s(t) \quad \text{and} \quad f_s \cdot q_s(t) < q_s(t - \lambda). \tag{10}$$

4.2   Case Study

This section will through case studies validate the effectiveness of our model in characterizing the process of task handling and study the dynamics of task handling.

In this section, we will also set $m = 2$. To satisfy the constraints discussed above, we can set parameter functions $f_1(t)$ and $f_2(t)$ as follows:

$$f_1(t) = \begin{cases} p_6 \cdot h_1(t, \lambda)/q_1(t), & \text{if } h_1(t, \lambda) > 0, \\ 0, & \text{if } h_1(t, \lambda) \leq 0, \end{cases} \tag{11}$$

$$f_2(t) = \begin{cases} p_7 \cdot h_2(t, \lambda)/q_2(t), & \text{if } h_2(t, \lambda) > 0, \\ 0, & \text{if } h_2(t, \lambda) \leq 0, \end{cases} \tag{12}$$

where $h_1(t, \lambda) = min(q_1(t - \lambda), q_1(t))$, $h_2(t, \lambda) = min(q_2(t - \lambda), q_2(t))$, p_6 and p_7 can be functions with a range of $[0, 1]$ or constants in $[0, 1]$. In our later case studies, we set them as constants. Without loss of generality, we set $g(t) = 300$.

Case Study 3. *This case study aims at examining the effectiveness of our task handling model as well as the effect of time threshold T. The parameters are set as follows:* $S(0) = 1000$, $\tau = 0$, $\lambda = 30$, $p_1 = 0.1$, $p_2 = 0.1$, $p_3 = 0.1$, $p_4 = 0.3$, $p_5 = 0.2$, $p_6 = 0.1$, $p_7 = 0.2$, $q_0(0) = 100$, $q_1(0) = 100$, *and* $q_2(0) = 400$.

We can note from the results in Figure 3 that:

1. For all cases of time threshold T, their $q_0(t)$, $q_1(t)$, and $q_2(t)$ curves converge to zero finally. That means all tasks are handled and corresponding agents disappear at last. At the same time, all curves remain non-negative. These are necessary conditions of the effectiveness of our characterization on task handling.

2. In all cases of T, we can observe two stages on all curves: an increasing stage and a decreasing stage. Due to $T \neq 0$ and $\lambda \neq 0$, that means at the beginning T steps there are new tasks submitted, but no old tasks are finished, therefore all curves increase at this stage. After T and λ exceed the preconcerted values, because no new tasks are submitted and at the same time some old tasks are handled, all curves begin to decrease and finally converge to zero[4].

3. For different T, $q_0(t)$, $q_1(t)$, and $q_2(t)$ curves reach their peaks at different time. The larger the time threshold T, the later the time when the curves reach their peaks.

Remark 3. From Case Study 3 and some other case studies (not presented), we observe that: (1) All curves corresponding to the numbers of wandering agents and various agent teams converge to zero eventually. That means all agents automatically disappear finally as all tasks are handled. Meanwhile, all curves keep non-negative. These two points are necessary conditions of the effectiveness of our model; (2) Service time λ and parameters p_6 and p_7 mainly determine the speed of task handling: the smaller the service time λ, the faster the speed of task handling; the larger parameters p_6 and p_7, the faster the speed of task handling. These are consistent with the (physical) meanings of λ, p_6, and p_7; (3) Time threshold T only affects the total number of tasks handled. A larger T indicates there are more tasks to be handled; (4) The effects of time delay in the task handling model is the same as those in the task distribution model. It only linearly prolongs the time to handle all tasks.

---

[4] Note that all $q_2(t)$ curves have a minor decreasing stage during the first 4 steps. It is caused by the following reason: at the beginning, as compared with the numbers of wandering agents and agent teams of size one, there are relatively more agent teams of size two, therefore some agents will leave these teams to wander and these teams become teams of size one. This can also observed in Figure 1.

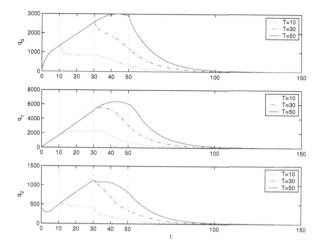

Fig. 3. Case Study 3: Task handling

5   Conclusions

Grid computing aims at sharing and integrating distributed computational resources and data resources. In grid computing, peer-to-peer grids are of special interest. In peer-to-peer grids, tasks are distributed to lots of grid nodes in a decentralized fashion. Facing such a large-scaled task allocation problem, this paper tried to provide a suitable computing mechanism and then examine the global dynamics of peer-to-peer grids. Specifically, we first presented an agent-based computing paradigm for peer-to-peer grids. We then provided two models to characterize two typical scenarios in this paradigm, i.e., task distribution scenario in short time intervals and task handling scenario in long time intervals. Through case studies, we validated the effectiveness of our models under different conditions, including initial agent distributions, time delay, etc. At the same time, we observed some key features of the dynamics of task distribution and handling.

Regarding the future work, we have the following two directions:

1. The paper addresses a special peer-to-peer grid computing scenario, where all grid nodes provide the same services and all tasks need the same services. In our future work, we need to relax the above restrictions on grid nodes and tasks so as to extend our model to a more general scenario, where both grid nodes and tasks can be heterogeneous.
2. In the paper, we analyzed our model through case studies. As the next step, we will develop a real platform so as to experimentally simulate and then validate our proposed computing paradigm as well as task distribution and task handling models.

References

1. Berman, F., Fox, G., Hey, T., eds.: Grid computing: making the global infrastructure a reality. John Wiley and Sons (2003)
2. Foster, I.: Internet computing and the emerging grid. Nature Web Matters (2000) http://www.nature.com/nature/webmatters/Grid/grid.html.
3. Foster, I., Kesselman, C., eds.: The Grid: Blueprint for a new computing infrastructure. Morgan Kaufman (1999)
4. Ge, Z., Figueiredo, D.R., Jaiswal, S., Kurose, J., Towsley, D.: Modeling peer-peer file sharing systems. In: IEEE INFOCOM 2003 – The Conference on Computer Communications. Volume 22. (2003) 2188–2198
5. Overeinder, B.J., Posthumus, E., Brazier, F.M.: Integrating peer-to-peer networking and computing in the agentscape framework. In: Proceedings of ICP2PC'02, Linkoping, Sweden (2002) 96–103
6. Gong, L.: JXTA: A network programming environment. IEEE Internet Computing **5** (2001) 88–95
7. Li, T.Y., Zhao, Z.G., You, S.Z.: A-peer: An agent platform integrating peer-to-peer network. In: Proceedings of CCGRID'03, Tokyo, Japan (2003) 614–617
8. Moro, G., Ouksel, A.M., Sartori, C.: Agents and peer-to-peer computing: a promising combination of paradigms. In: Proceedings of the 1st International Workshop on Agents and Peer-to-Peer Computing. Volume 2530., Springer (2003) 1–14
9. Lerman, K., Shehory, O.: Coalition formation for large-scale electronic markets. In: Proceedings of ICMAS 2000. (2000) 167–174
10. Lerman, K., Galstyan, A., Martinoli, A., Ijspeert, A.J.: A macroscopic analytical model of collaboration in distributed robotic systems. Artificial Life **7** (2001) 375–393
11. Liu, J., Jin, X., Wang, Y.: Agent-based load balancing on homogeneous minigrids: Macroscopic modeling and characterization. IEEE Transactions on Parallel and Distributed Systems (in press) (2004)
12. Wang, Y., Liu, J., Jin, X.: Modeling agent-based load balancing with time delays. In: Proceedings of IAT-2003, Halifax, Canada (2003) 189–195

Peer-to-Peer Computing in Distributed Hash Table Models Using a Consistent Hashing Extension for Access-Intensive Keys

Arnaud Dury

Centre de Recherche en Informatique de Montral,
550 rue Sherbrooke Ouest,
Montral, H3A 1B9, Qubec, Canada
Arnaud.Dury@crim.ca

Abstract. Classical distributed computing projects generally use a specialized client/server model. Recent approaches, such as BOINC, favor instead the development of distributed computing platforms, relying on a generic client/server model. We propose a fully decentralized computing model, considering all participant as peers that can submit personalized computing tasks to any number of other peers currently offering their services, listed in a peer directory. Our model is built upon Chord, a particular Distributed Hash Table. Chord allows load balancing of the number of keys per node, but offers no way to balance the bandwidth load of a frequently accessed key, such as a peer directory. Our model extends Chord with load-balancing of those access-intensive keys. We present a modelization of the bandwidth and storage costs of our model and experimental performance results using a variable number of peers, tasks, tasks time, and a variable ratio of contributors and solicitors roles among peers.

1 Introduction

Distributed computing is a tool used in a growing number of research fields: mathematics [1][2], biology [3], radio-signal analysis [4], protein folding [5], [6], genome analysis [7], [8], meteorological previsions [9], crypto-analysis [10] and others. These works are performed in a massively distributed manner, using idle time from computers of generous contributors over Internet. But for their popularity, the number of participants is still limited by the absence of tangible rewards for the contributors, and the barriers of entry for any new project are high. We present in this paper a Peer-to-Peer model for distributed computing addressing these two issues, based on distributed directories over a Distributed Hash Table (DHT). We propose a new solution for the handling of the "hot spots"[1] such directories generate, and for which classical caching algorithms in P2P systems are not applicable. We present a theoretical modelization of the

---

[1] Hot spots are keys in a DHT so frequently accessed that they introduce network congestion for the nodes responsible for them.

G. Moro, S. Bergamaschi, and K. Aberer (Eds.): AP2PC 2004, LNAI 3601, pp. 185–192, 2005.
© Springer-Verlag Berlin Heidelberg 2005

bandwidth and storage cost or our model, and we relate this modelization to experimental results. The first part introduce our model, which extends the consistent hashing used in Distributed Hash Table to balance the load of keys that are both frequently read and frequently modified. We present a study of performance of our model, by modelization and experiments, and we conclude on the possible use and future extensions.

2   Network Efficient Nodes Discovery in a Distributed Hash Table

The two main concepts of our model are the indexing of idle contributors nodes and solicitors nodes in two distributed directories over a Distributed Hash Table (DHT), and the use of the same DHT to store computed results as soon as they are produced, if their originator is not connected. We developed a distributed directory model, that offers load balancing of the bandwidth among the nodes, and provides exhaustive answers to queries while preventing the flooding of the network. We chose the DHT model (Chord, more precisely) over more classical approaches such as Gnutella [11], Napster [12] or FreeNet [13], because DHT models offer provable communication costs, provable stability under nodes join and leave and support for both Read and Write operations. A distributed computing system can have at times an excess of contributors, and at other times a excess of solicitors. Our model uses two distributed directories, one for solicitors and one for contributors. When a contributor node joins the network, it checks for existing solicitors in the solicitors directory. If any are found, it will contact one of them, chosen randomly, to collect units to process. If none are found, the contributor will register itself to the contributor directory. The same principle applies when a solicitor joins the network. In a DHT, an information is stored under a key, on a certain node. The bandwidth of the node storing the directory will quickly become saturated, because it will receive all the requests made by all the nodes of the system. The load balancing present in Chord is a balancing of the number of keys among the nodes, but not a balancing of the load access to a particular popular key. While caching extensions to Peer-to-Peer systems (such as [14]) have been proposed to solve this problem of "hot spots", or highly accessed keys, they are not applicable to directories: in a period of high activity, nodes may enter and leave them frequently, and caching is inefficient in the case of frequently modified data. We present an alternative and more efficient method, based on an extension of consistent hashing.

3   Segmentation of an Access-Intensive Key over a Consistent Set of Linked Nodes

Chord associate to each key K_i a hash code H_i using a function $Hash(K_i) = H_i$. The node ω_i responsible for K_i is the node whose identifier is the closest to H_i when proceeding in a clockwise manner from H_i on the identifier ring.

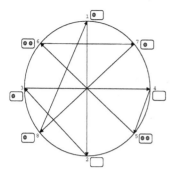 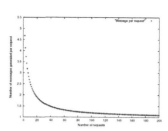

Fig. 1. Segments choice strategy (left) - Number of messages per request (right)

We propose in this paper to extend this hashing from $Hash(K_i) = H_i$ to $Hash(K_i) = \{H_j : H_j = \phi(Hash(K_i), j), 1 \leq j \leq k\}$. We construct ϕ such as $\phi(Hash(K_i), 1) = Hash(K_i)$, and in such a way that the values returned by ϕ can be computed independently by any node of the system. The function ϕ should not need any meta-information stored in the DHT, because the node storing this meta-information would become the new saturated node. We use the name "segmented list" for the set of nodes $\{\omega_1, \omega_2, ..., \omega_k\}$ that are responsible for the hash code $H_1, H_2, ..., H_k$, the name "segment" for each of these nodes to differentiate them from others nodes of the system, and the name "segment population" to define the number of identifiers stored in a segment. Each of the segments is linked to the next one, in a circular manner 1. A nodes directory is thus a named, segmented list. The choice of the ϕ function embodies the strategy for the key distribution. The strategy we choose tries to avoid the use of nodes already responsible for the replication of another part of the same list by allocating further segments of the list to segments currently as far as possible from the existing ones (see 1). The function ϕ is thus only based on the key hash code (defining the location of the first segment location on the ring) and the number of segments (or "hops"), and can be computed independently by any node. A node willing to register randomly chooses a segment number from 1 to the estimated bound of the number of segment. *This random choice will make nodes spread the load on the directory key over the sets of segments responsible for it.* The bound will be computed using the number of nodes per segment, which is a parameter of the system, and the estimated total number of nodes in the system, which is impossible to compute precisely, but for which we can obtain a good estimation. In a Chord ring, each node maintains an index table of successors at successive power of two from them. Each node ω_i compute the angular density of nodes on the ring, using data from its index table. Assume l is the number of entries in the index table of ω_i, $idx_i[l]$ is the last entry, and $\Theta(\omega_a, \omega_b)$ is the angle from ω_a to ω_b on the ring. The angular density d is calculated as $\frac{2^l}{\Theta(\omega_a, idx_a[l])}$. The total number of nodes is then estimated by $T = 360 \times d = \frac{360 \times 2^l}{\Theta(\omega_a, idx_a[l])}$. We note β the maximal segment population. The total number of segments is thus $\frac{T}{\beta}$. A registration request to a directory of name K_i is sent to the node responsible for the hash code $\phi(Hash(K_i), j)$ where j is chosen randomly in the $[1, \frac{T}{\beta}]$ interval. The node receiving this request will register the node if its local population is under β, and will confirm to the subscribing node that its registration is done. If the node has reached its maximal population, it will forward the request to the next node in the chain, until a suitable one is found. Directory lookups follow a similar method, while directory removal is a direct communication to the corresponding segment.

4 Theoretical Modelization

We introduce a modelization of memory and bandwidth costs associated with two models: the DHT storing the whole index in one node, and our model using a segmented list that can be accessed at any segment. T is the number of nodes in the system. Each node can act as a contributor or a solicitor whenever it wants. L is the number of nodes that are solicitors (provider of works units) $(L \leq N)$. Δ is the average time of computation for a work unit in seconds. U is the total number of work units to compute. B is the maximal upload bandwidth consumed. S is the size of request and reply packet. We assume $S = 1500$ bytes. β is the maximal population of a segment. $1 \leq \beta \leq T$. Assuming that $U \gg T$, we compute the bounds of the memory and bandwidth imposed to the node storing the directory for a classical DHT, during a period of $d = 1$ seconds. This node will store every identifier of the system. We assume these identifiers consume m bytes each. The memory consumed is thus $M = \frac{T*m}{1024}$Kb. The node responsible for the key will also answer directory look-ups. We compute the upload bandwidth (bandwidth used for replying to request) for this node: $B = d * \frac{T}{\Delta} * S$ bytes per second. Such a model can keep with at most : $T = \frac{B*\Delta}{d*S}$ nodes. Assuming a connexion with a upload bandwidth limit of 512 kbit/s (optimistic upper limit of most broadband ADSL lines), and assuming $\Delta = 60s$ on average, the maximal supported number of nodes is slightly over 2600.

4.1 Linked List

Using a linked list of nodes, the first ones of which would be the one directly responsible for the key, would be useless: the memory constraints for each node would be lessened to $M = \frac{T*m}{1024*\frac{T}{\beta}} = \frac{m*\beta}{1024}$Kb, but the bandwidth load would stay the same for the head of the list. The memory requirements for our directory model are the same than in the linked list case. There is thus no theoretical limit to the number of nodes that the system can accommodate, memory-wise, because in an extreme situation each node may be responsible for as many or as few identifiers as we choose.

4.2 Worst, Best and Average Cases

We study the new bandwidth requirements, under worst-case, best-case and average-case situation. In the worst case scenario, new nodes willing to act as solicitors arrive constantly while all the contributors are occupied: no request can be satisfied, and each one has to go through all the nodes of the list. But in this case, each solicitor will only generate one request before subscribing itself to the solicitors list, and entering a wait state. In most conditions, the arrival of new solicitors can be assumed to be spread over time. So even in this worst case, the system would probably stay efficient due to its use of a double directory structure, avoiding too many active polling for resources. In the best case scenario, the system is in the following state: no segment of the list ever becomes empty, due to a sufficiently large number of contributors. In this situation, each contributor request is answered immediately by the first segment receiving it. The bandwidth generated by the request is minimal, and the system can expand indefinitely. In the average case scenarion, the system stays between the two previous extremes. A number of contributors register as idle each second, and a number of solicitors send requests each second. We model the equilibrium case, with an average equal number of registration and request per second. We compute the total consumed bandwidth, in number of messages, to answer R simultaneous requests from a segmented list containing R identifiers. We assume that the R requests will be answered before any new contributor arrives. Let α be the number of segments in the chain: $\alpha = \frac{T}{\beta}$. A solicitor requesting a contributor identifier send its request to any node of the segmented chain, and this message is forwarded until an answer is found. The total number of messages produced depends on the density of contributors identifiers available in the chain. We consider a request reaching a randomized segment. To simplify the model we restrict ourselves to the case where $R \leq \beta$. For one identifier, the chance to be absent from the first segment reached by the request is $\psi_1 = \frac{\alpha-1}{\alpha}$. The chance to be absent from the segment i of the list, knowing that the $i - 1$ previous segments are empty is $\psi_i = \frac{\alpha-i}{\alpha-i+1}$. The segment i is empty if all of the R identifiers are absent. Using the assumption $R \leq \beta$, there is no dependency between the location of an identifier and the locations of the others. The probability $\Psi_i(R)$ that the first i segments are empty is thus: $\Psi_i(R) = \psi_i^R = (\frac{\alpha-i}{\alpha-i+1})^R$. We compute now the chance to discover at least one identifier in the segment i after having traversed the first $i - 1$ empty segments as : $\lambda_i(R) = 1 - \Psi_i(R) = 1 - (\frac{\alpha-i}{\alpha-i+1})^R$. Now we can compute the average number of requests messages generated by one request for a contributor identifier. The request have a probability $\lambda_1(R)$ to be satisfied by the first segment asked, thus generating only one request message (the initial request itself). If unsuccessful (with a probability $1 - \lambda_1(R)$), the request then have a probability $\lambda_2(R)$ to be satisfied by the second segment, thus generating two requests messages, the first request, and its retransmission from the first segment asked to the segment one. The probability for the request to be satisfied after $s \leq \alpha$ steps is thus $Satis(s) = \lambda_s(R) \times \prod_{\delta=1}^{s-1}(1 - \lambda_\delta(R))$. We assume than the distribution of identifiers stays homogeneous after each request has been satisfied. This assumption implies that there is a redistribution of the identifiers after each request is processed, which is not the case. We are overestimating the homogeneous spread of identifiers. We compute thus M, a lower bound of the average number of messages needed to answer **one request**, while there are R identifiers left in the chain.

$$M(R) = \sum_{s=1}^{\alpha} s \times Satis(s) \tag{1}$$

$$= \sum_{s=1}^{\alpha} \left(s \times \lambda_s(R) \times \prod_{\delta=1}^{s-1} (1 - \lambda_\delta(R)) \right) \tag{2}$$

$$= \sum_{s=1}^{\alpha} \left(s \times (1 - (\frac{\alpha - s}{\alpha - s + 1})^R) \times \prod_{\delta=1}^{s-1} (\frac{\alpha - \delta}{\alpha - \delta + 1})^R \right) \tag{3}$$

The *total* number of messages produced to answer R requests is thus $\sum_{i=1}^{i=R} M(i)$. We show on the figure 1 the predicted average number of messages needed to process one request, over a variable number R of requests for a list of R identifiers present in the list of contributors of 10 segments, and a maximal segment population of 20. This scenario is our equilibrium state scenario describe previously. Our model predicts an interesting property: the higher the number of requests received in an equilibrium state, the lower the average number of messages needed to answer each one is, given a fixed number of segments and a fixed segment population. We will show how the experimental results confirm this modelization.

5   Experimental Results

We use a Network Simulator we built in Java, that allows us to parametrize the connexion speed between each pair of machines. We coded our DHT model on top of this simulator. All the computations at each node are simulated (time taken and results size are parametrized). There is no packet loss, congestion only lead to longer delay between machines. All nodes have the same computing power, all units take the same time to be computed and all nodes have the same bandwidth, that of a high-speed ADSL connexion. We show the experimental results obtained with our model, using a variable population of nodes, units and computation time and a variable mix of contributors versus solicitors. The upper-left figure 2 shows the acceleration factor with a variable number of nodes, from which only one acts also a solicitor, $\Delta = 3000$ seconds and $U = 3000$. The upper left figure 2 shows the acceleration factor with $T = 300$ nodes, $U = 3000$ units and a variable unit-time. These results show that the acceleration factor is near the number of nodes when the unit computation time is high, which makes sense as the P2P infrastructure introduces delays of its own, which have more impact with short computation times tasks. Acceleration factor is higher then the number of units is a multiple of the number of nodes[2].

---

[2] 301 units computed with 300 nodes will take Δ more seconds than 300 units computed on the same number of nodes.

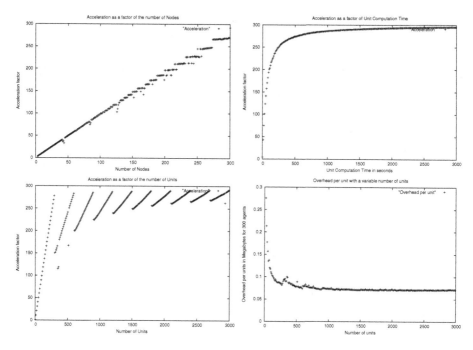

Fig. 2. Acceleration factor. Overhead of the system over a variable number of units.

On the middle-right figure 2, we see that the simulation confirms our theoretical model: the higher the number of units, the lesser the communication cost per unit is, keeping the number of nodes the same. The cost decreases as long as there is more nodes than units, and then stabilizes, because solicitors stop emiting new requests and start using their own directory.

6   Conclusion and Future Works

We introduced a new model for the efficient distribution of a directory model over a DHT, providing bandwidth load-balancing for the frequently accessed directories entries. A permanent connexion is not required to collect all the returning results. While our current implementation runs on simulation, the real implementation will use the Java Virtual Machine as its security and code mobility layer. Modelization indicate that the bandwidth cost per computed units is lowered when the number of units increases, which we verified in our implementation. We are now working on a real implementation of this model, offering a generic API to allow anyone to interface their code with our P2P network. We will implement research applications such as a distributed version of the Spin model-checker (see [15]), and a distributed version of a genetic algorithm library. A first possible improvement of our model is inter-segments communications. Each node will communicate with its two neighbors in two cases: node switch from empty state to not-empty state, and node switch from full state to not-full state. Propagation of information from node to node will allow them to know the closest node with free space, or with an identifier of an idle contributor. This allows quicker registration and request phases, at the cost of new communications messages between neighboring nodes. Another improvement will be to take into account meta-data for each node, such as the computing power of the node, in the request phase. This will allow each solicitor node to choose the best available contributor, and to implement a distributed scheduling algorithm. Pre-caching on the DHT of future units to distribute is also considered, when upload bandwidth is available. The bandwidth bottleneck that occurs when too many contributors return their results and ask for new units at the same time will be alleviated.

References

1. Odlyzko, A.: Zeros of the riemann zeta function: Conjectures and computations. In: Foundations of Computational Mathematics conference, Minnesota (2002)
2. http://mersenne.org (1996)
3. Loewe, L.: evolution@home: Experiences with work units that span more than 7 orders of magnitude in computational complexity. In: 2nd International Workshop on Global and Peer-to-Peer Computing on Large Scale Distributed Systems, IEEE Computer Society (2002) 425–431
4. University of California, B. (http://setiathome.berkeley.edu/)
5. Zagrovic, B., Snow, C.D., Shirts, M.R., Pande, V.S.: Simulation of folding of a small alpha-helical protein in atomistic detail using worldwide distributed computing. Journal of Molecular Biology (2002)
6. et al., V.P.: Atomistic protein folding simulations on the submillisecond timescale using worldwide distributed computing. Biopolymers (2002)
7. SM, L., A, G., JR, D., VS, P.: Increased detection of structural templates using alignments of designed sequences. In: Proteins: Structure, Function, and Genetics. (2003) 390–396
8. Larson, S.M., Snow, C.D., Shirts, M., Pande, V.S.: Folding@home and genome@home: Using distributed computing to tackle previously intractable problems in computational biology. In Grant, R., ed.: Computational Genomics, Horizon Press (2002)
9. of Oxford, U., the Rutherford Appleton Lab, University, T.O. http://www.climateprediction.net (2003)
10. (http://www.distributed.net/)
11. OpenSource. (http://www.gnutella.com)
12. (http://www.napster.com)
13. Clarke, I., Miller, S., Hong, T., Sandberg, O., Wiley, B.: Protecting free expression online with freenet (2002)
14. Naor, M., Wieder, U.: Novel architecture for p2p applications: the continuous-discrete approach. In ACM, ed.: SPAA, San Diego (2003)
15. Barnat, J., Brim, L., Stribrna, J.: Distributed ltl model-checking in spin. In: Lecture Notes in Computer Science. (2001) 200–216

A Practical Peer-Performance-Aware DHT

Yan Tang, Zhengguo Hu, Yang Zhang, Lin Zhang, and Changquan Ai

School of Computer, Northwestern Polytechnical University, Xi'an, P.R. China
{tangyan, zhangy}@co-think.com
{zghu, geniuslinda}@sina.com
shrex_acq@hotmail.com

Abstract. How to Build an efficient Distributed Hash Table (DHT) is a fundamental issue in Peer-to-Peer research field. Previous solutions ignore the heterogeneity of the large scale network. However, in practice, the fact is that the resource held by each peer in the Internet is extremely diverse. And the the willing to share local resources of each peer is also diverse. Therefore, the contribution for the system of a peer should depend on the resources it holds or how many resources it want to share, and should not be uniform. In this paper, we propose a Peer-Performance-Aware Distributed Hash Table (PPADHT) which aims to exploit the heterogeneity. It takes the performance difference of peers into consideration to construct a dynamic variation of wrapped butterfly to achieve the goal. We also show how to optimize the performance of PPADHT in the view of hop counts by random graphs. Our simulation results show that the average lookup hop counts of the PPADHT is approximately a log scale with constant out degrees. And it can achieve loadbalance in two ways: both the document load and message routing load, without introducing any additional load on the peer. Here, the load balance means the load is proportion to the performance of peer.

1 Introduction

Distributed Hash Table (DHT) is now a widely studied Peer-to-Peer (P2P) infrastructure. It is outstanding because of the following advantages: it is purely distributed, extensible,accurate, and balances load. Researchers now study the DHT under the assumption that each peer joined the overlay network has same performance (the CPU performance, storage, bandwidth and so on). This assumption is convenient while studying traditional server clusters. But in the application environment of a P2P infrastructure, the assumption is not feasible. As reported in [1], extreme heterogeneity of peer performance exists. Ratnasamy[2] proposed the open question: Can one redesign DHT routing algorithms to exploit heterogeneity? With this in mind and more, we present a method to represent the synthetic performance of the peers joining the P2P system in the overlay construction process that may exhibit better properties than one can find when building the overlay network without considering the real environment.

The remainder of this paper is organized as follows: Section 2 reviews the related work; section 3 presents the basic assumptions and motivation for our work;

G. Moro, S. Bergamaschi, and K. Aberer (Eds.): AP2PC 2004, LNAI 3601, pp. 193–200, 2005.

section 4 shows the details of the PPADHT; section 5 presents the simulation results; and section 6 summarizes the paper.

2   Related Work

Currently, all the DHTs are built based on the flat one-dimension space and provide load balancing based on the assumption that the performance of peers is equivalent. According to the report of Saroiu et al.[1], the bandwidth, latency and availability of peers are diverse from three to five orders of magnitude. And it is noticed by many researchers. The work of Kwon et al.[3] provides a hierarchical way to capture the heterogeneity of the Internet. The work of Yingwu[4] bases on virtual server, and utilizes collected proximity information to guide the migration of virtual server. Recently, the work in [5] utilizes the heterogeneity of nodes to modify the one-hop overlay [6]. They use level to present the bandwidth of the nodes in the overlay and nodes at different level hold different size of routing table and maintenance overhead. We utilize the heterogeneity in a different way.

3   Assumptions and Motivation

Previous works have the following basic assumptions:

Each peer in the P2P network has similar performance.
The documents are distributed in the overlay space randomly and evenly.
Any of the peers can freely connect with each other.

In practice, it is possible to make hundreds or even thousands of server peers which have likely performance to be organized into an overlay. When the number of peers achieves a large scale like the Internet, the heterogeneity becomes a natural feature. Our work bases on the following three basic assumptions:

The performance of each peer in the P2P network satisfies the power-law distribution.
The documents distributed in the overlay space randomly and evenly.
The popularity of documents also satisfies the power-law distribution.
Any of the peers can freely connect with other peers.

Our goal is to exploit the heterogeneity in the large scale P2P infrastructure and to try to build a more practical large scale DHT system by considering more complex conditions that approach the reality of the current Internet's behavior. And the power-law distribution here, is truncated as the highest level hold all the rest probabilities.

4   The Peer Performance Aware DHT

4.1   Overview

The basic idea of our Peer Performance Aware DHT (PPADHT) is to combine a two-dimension space into the flat one-dimension overlay space. One of the two dimensions represents the performance of a peer; the other is generated by Hash function. The fundamental topology utilized in the PPADHT is wrapped butterfly network and the variant of it. We borrowed most mechanisms from the chord project, including the successor list, and stabilization.

4.2 Overlay Space

The overlay space of PPADHT can be split into two parts. The high part is an $\log(k)$ bits number represents the performance of a peer, and it is named as level of a peer. The low part is a k bits number named as the row identifier of a peer. It is generated by SHA1 algorithms. They are combined to be one $k + \log(k)$ bits identifier. So the whole overlay space is $[0, 2^{k+\log(k)} - 1]$. Given any two identifiers X and Y, The clockwise distance and real distance of them can be calculated by these formulas:

$$d(X, Y) = (Y - X + 2^{k+\log(k)}) \% 2^{k+\log(k)}$$

$$rd(X, Y) = \min(d(X, Y), d(Y, X))$$

As in Chord, each peer and each document has a unique identifier. The identifier of a document is called key of that document. A peer holds the documents that have the key between it and its immediate predecessor in the overlay space. Unless for the last peer in the key space, it will hold the documents with key from its immediate predecessor to $2^{k+\log(k)} - 1$. It looks unfair for the last peer, but we assume it have better performance, it should have enough capability to do so. And this problem can be solved by the virtual server technology, we will use it to balance the load between peers in the same level.

4.3 Routing Protocol

Routing Table. The routing table of a peer in the PPADHT contains the edges that point to the butterfly edges of the peer, the immediate predecessor, successor lists and an edge that point to the peer that has same row identifier with it but at level $((l-1) + k) \% k$ named ancestor. As the performance of such structure in the view of hop counts is not very good, we also add some random edges in the out-degree which will be discussed in section 5.1.

Wrapped butterfly network is a variant of butterfly network. Let us write l for the level of a peer, $e_0 e_1 ... e_k$ for the row identifier of a peer, and then a peer in a wrapped butterfly network could be uniquely represented as $< l, e_0 e_1 ... e_k >$. Suppose totally there are k levels (from level 0 to level $k-1$) in the wrapped butterfly network, then each level has 2^k peers. A peer $< l, e_0 e_1 ... e_k >$ has two out degrees pointing to peer $< (l+1) \% k, e_0 e_1 ... e_l ... e_k >$ and $< (l+1) \% k, e_0 e_1 ... \overline{e_l} ... e_k >$ respectively. The diameter of the wrapped butterfly network is $\lceil \frac{3}{2} k \rceil$. Please refer to [7] for the detailed proof.

We have three reasons to select the wrapped butterfly network as the fundamental structure. Firstly, it is a network born with two dimensions. It is easy

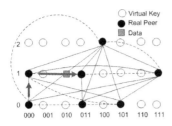
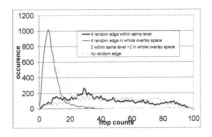

Fig. 1. A simple example of a lookup

Fig. 2. The effect of deferent kinds of random edges on the hop counts

for us to use one of the dimensions to represent the performance of a peer. Secondly, it has low diameter. Thirdly, its degree is constant, which means it is degree optimal.

Next Hop Decision. Two kinds of next hop decision can be applied. One comes from the simple package routing algorithm [7] of wrapped butterfly network. The other is the greedy algorithm, just to take the edge which is nearest to the target. The former can achieve $O(\log(n))$ hop counts which high probability, but k will be a big constant, it is not really good in practice. The later can only achieve $O(n)$ hop counts, but with the random edges, it can achieve better result than the former method. The reason we focus on the greedy algorithm is because it is simple and robust. We will discuss it in the section 4.6. For a given key, find the peer, which is the nearest one to the key in the routing table and forward the message to it. The distance is the real distance of two peers. If the peer is the current peer itself and it does not hold the key, the message will be forwarded to the immediate successor. The process terminate till the message arrives to a peer that is responsible for the key. Figure 1 shows a simple example of the greedy lookup process (we do not present the random edges in this figure).

Peer Joining and Leaving. A new peer that wants to join the overlay must know a peer that is already in the overlay. The bootstrap process is the same as chord except the way to establish finger table. We just need to locate two butterfly edges, an ancestor edge and several random edges. And the immediate predecessor and successor of the new joined peer will notify the peers pointing to them by butterfly edges and ancestor edges. The cost of join operation is just linearly related with the hop counts of a lookup. We will show in our experiments that the hop counts of a lookup are logarithmically increasing with the number of peers and the worst case is well bounded. So the cost of join operation is also logarithmically increasing, which means the overlay will have good scalability.

A peer leaving the overlay should transfer the documents it holds to its successor and notify all the peers that pointing to it. The cost of peer leaving can be ignored as it will not perform any lookups.

To Shorten the Hop Counts. The greedy routing algorithm can only achieve $O(n)$ hop counts for a lookup. Recall that random graphs always have low diameter, we try to improve this by embed random graph into the overlay. We use the

MT random generator [8] to randomly generate the identifiers and let the peer establish edges pointing to those peers that are responsible for these identifiers. And this method is shown to be very efficient in our simulation.

Three ways exist to add random edges for a peer. We can generate the random identifiers in the whole overlay space, or generate it at the same level of the peer, or add both kinds of random edges. To generate random identifier in the whole overlay space means, the in-degree of peers will follow the power-law distribution. As we assumed, peers distributed among levels follow the power-law distribution. Then the overlay space hold by different peers also follows the power-law distribution. And the random-edge of every peer will point to the key randomly and evenly distributed in the overlay space. So the peers hold large proportion of the overlay space will have more in-degrees. And we believe the power-law distribution of in-degree edges is the reason why the method is so efficient in this way.

4.4   Discussion

Level of a Peer. Let us write B for network bandwidth, T for the rate of uptime and downtime of the peer stay in the network, C for CPU performance, M for memory capacity, S for storage capacity, then level of a peer is a function L of B, T, C, M and S, say, $level = $ L (B, T, C, M, S). Here, B and T play an important role for level because B is the main concern of the peers in the Internet, and T presents the stability of the peer and it is based on the statistic of the past behavior of the peer. We do not want those peers with high bandwidth but join and leave the overlay frequently to be placed at high level as the cost of leave and join operation for a high level peer is very high. Here we ignore the influence made by C, M and S to make the above formula easier, $level = $ L (B, T). The influence of the performance of a peer can only be evaluated with a large scale test-bed. We are planning for a new global wide application with our PPADHT to perform further analysis on the level selection of a peer.

Load Balance. We discuss the load balance of the PPADHT in two aspects: document number hold by each peer and the message transfer load of each peer. The basic assumption of our discussion is peers distributed among levels satisfy the power law distribution $p(x) = (x + \alpha)^{-\beta}$. We plus a constant α on x to avoid level 0 hold all of the probability.

We assume that there are N documents randomly and evenly distributed among all the peers. Peers present at level i would be within $np(i) + np(i)(1 - p(i))$ with high probability and $np(i)$ on average case. The document distributed at each level is within $\frac{N(2k-1)}{k^2}$ with high probability and N/k on average. Then peers in level i will responsible for at most $\frac{N}{n} \frac{2k-1}{k^2}(i + \alpha)^{2\beta}$ documents with high probability and $\frac{N}{n} \frac{(i+\alpha)^\beta}{k}$ on average. Where n, N, α, β and k are the same for all peers. Such that the document hold by each peer at different level will just have relationship with the level of it. And we also calculate the level according to the peer performance. So we can achieve document distribution load balance in the view that high capacity peer contributes more for the network. And if users do not want to contribute so much, they can adjust the load by adjusting the level. We can see obviously from the formulas that the document load is exponentially increasing as the level increasing.

To discuss the load balance for message transfer, we assume that to lookup a document is behaved in a random and even distribution way. As the probability for a peer to present at high level is fewer and fewer, higher level peers will have more in bound linkages. The chance to take a high level peer as a hop is larger. At the same time, higher level peers hold more documents, so the chance to access them is larger. As we assumed, the popularity of documents also satisfies the power law distribution. Things got to be change. It can not be solved by the structure itself. We will solve it by the replication protocol in the future work.

5   Evaluation

We build a simulator using Java language to evaluate the protocols of PPADHT. The experiments were done on a blade2000 workstation with 2G memory and dual ultrasparcIII 1.2G CPU. In our experiment, we construct the wrapped butterfly network with totally 256 levels, and the peers generated in different levels satisfy the power law distribution $p(x) = (x + \alpha)^{-\beta}$ with parameter $\beta = 2.07$ and $\alpha = 5$. We do not use the successor list and stabilization mechanism described in Chord, as we think they will have great influence on the property of the overlay, especially with small number of nodes.

5.1   Hop Counts

As we have discussed, only the last hop counts of the greedy algorithm can not be well bounded while the lookup get to a peer that is at the same level of the target identifier. So the way to generate the random identifier at the same level of a peer can make the lookup well bounded. But the first method will achieve really low diameter in the whole overlay. To make a tradeoff, add both edge should be good. The hop counts without random edge is similar to have random edges at the same level, but the worst case is worse than it. The Figure 2 shows it. If without butterfly edges, the greedy lookup will cause many loops in the overlay and messages can not be routed to the target correctly, so we do not show the experiment result of it. Figure 3 shows the average hop counts increasing as the number of peer increasing. The parameter for experiment is 4 random edges on same level and 4 random edges in whole overlay space, 2^{10} to 2^{15} peers, and perform 10000 random lookup. It increases linearly, so we can say that the average hop counts increases logarithmically.

5.2   Load Distribution

Document Distribution. We did our experiment on document distribution under this configuration to show that our PPADHT has the ability balance the load of document distribution among levels. The parameter for the experiment

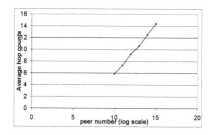

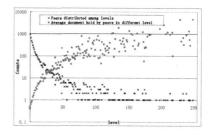

Fig. 3. Average hop counts as a function of logarithmic of peer number

Fig. 4. Peers distributed among levels VS. average count of documents hold by peers at different level

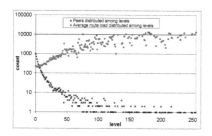

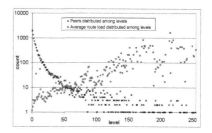

Fig. 5. Peer distributed among levels VS. average random lookup route load of peers at different level

Fig. 6. Peer distributed among levels VS. average power-law lookup load of peers at different level

is 5000 peers, 100000 documents randomly distributed. Figure 4 shows relation of distribution of peers among different levels and the average document load of peers in different level. From it, we can see that the count of document hold by a peer increases as the level increases, which means it can match the distribution of the peers distributed in different levels. Obviously the load of higher level peers differs a lot, this is caused by some levels do not have any peers, so all the load of those levels was put on the immediate succeed peer.

Message Routing Cost. To show the distribution of the load for message transfer, we did an experiment under this configuration: 10000 peers, 100000 documents, 200000 random lookups. Figure 5 shows the distribution of the load to route message. Also we can have the result that as the level of peer increases, it contributes more for the overlay.Figure 6 shows the relation of peer distribution among different levels and average route load of peer at different levels while the popularity of documents satisfies the power law distribution. It is clearly that some peers at lower level hold too much routing loads. We are scheduling to solve this problem by replication algorithms.

6   Conclusion and Future Work

In this paper, we present the PPADHT. It utilizes the wrapped butterfly structure to construct an overlay network, and takes the extremely performance diversity of Internet hosts into consideration, so as to provide load balance in the practice view which means higher performance peer should contribute more for the system. Simulation shows our approach can balance the load from two aspects, the data distribution and the message transfer. The idea to consider the distinguished peer performance while constructing overlay network should be feasible to improve the load balance and other properties of the DHT systems. We will continue to working on it. Simulation results also shows that the hop counts for message route in PPADHT can achieve logarithmically increase as peer number increases.

However, there are limitations of this approach. We have noticed that when the document popularity satisfies the power-law distribution, the message route load do not achieve a good state as some peers at lower level hold too much loads, much of the higher level peers does not contribute enough as they are assume to do. In the future, we schedule to solve this problem by working on the replication algorithms.

References

1. Saroiu, S., G.K., Gribble, S.: A measurement study of peer-to-peer file sharing systems. In: Proceedings of Multimedia Conferencing and Networking. (2002) 156–170
2. Ratnasamy, S., Shenker, S., Stoica, I.: Routing algorithms for dhts: Some open questions. (In: Proceedings of 1st International Workshop on Peer-to-Peer Systems)
3. Gisik Kwon, K.D.R.: An efficient peer-to-peer file sharing exploiting hierarchy and asymmetry. (In: proceedings of SAINT 2003)
4. Yingwu Zhu, Yiming Hu, E.: Proximity-aware load balancing for structured p2p systems. In: Proceedings of P2P'03. (2003)
5. Jingfeng Hu, Ming Li, N.N.W.Z.: Smartboa: Constructing p2p overlay network in the heterogeneous internet using irregular routing tables. (In: proceedings of iptps'04)
6. Anjali Gupta, Barbara Liskov, R.R.: One hop lookups for peer-to-peer overlays. (In: proceedings of HOTOS IX)
7. David S.L. Wei, F.P.M.I., Naik, K.: Isomorphism of degree four cayley graph and wrapped butterfly and their optimal permutation routing algorithm. (IEEE TRANSACTIONS ON PARALLEL AND DISTRIBUTED SYSTEMS)
8. Matsumoto, M., Nishimura, T.: Mersenne twister: A 623-dimensionally equidistributed uniform pseudorandom number generator. ACM Trans. on Modeling and Computer Simulation **8** (1998) 3–30

Peer-to-Peer Data Lookup for Multi-agent Systems

Michael Thomas and William Regli

Department of Computer Science, College of Engineering,
Drexel University, Philadelphia, PA 19104-2875
{mst34, regli}@drexel.edu

Abstract. Systems such as Napster and Gnutella demonstrated the potential of peer-to-peer data sharing. Similar schemes have been used to provide solutions that ensure information availability, survivability and reliability. Current techniques based on Distributed Hash Tables (DHTs) promise scalable solutions for efficient lookup when data is distributed across large networks.

This paper considers how to adapt DHTs for use with multi-agent systems, with a goal of supporting distributed data storage and lookup on resource-constrained devices operating on dynamic networks. In existing DHTs, the network and the data are assumed to be static. In our context, sets of mobile agents manage the data.

We present a multi-agent approach for building CAN-based DHTs. DHT access is provided through a DHT Agent Service. An extension of the standard CAN lookup algorithm is presented which allows more efficient index maintenance for highly mobile agents. Empirical results verify that the agent-based CAN achieves the expected scalability.

1 Introduction

Mobile, intelligent agents have become a widely accepted method of developing distributed applications, due to both the flexibility of the agent-based programming paradigm and the robustness provided by mobile agent architectures. This is particularly important in dynamic, resource-constrained environments, such as that posed by a group of soldiers carrying PDAs on an ad hoc network. In such an environment, an agent-based system can adapt to the loss of a network node by changing the migration paths of agents around that node to remaing network nodes.

The simplest agent-based systems include individual, independent agents; however, the capabilities of single, independent agents operating in isolation can only go so far. Multi-agent systems, however, require mechanisms for the various agents to communicate and share data and knowledge. This might take the form of shared workspaces, blackboards, service discovery, or data exchange/lookup mechanisms between agents. For example, if an agent requires a certain piece of data to perform its task, it can search or broadcast to the network to find out if some other agent in the system has acquired that data and can provide it to the searching agent. This can become problematic and inefficient in large systems and in complex, dynamic network domains. Simple broadcast or flooding approaches, while suitable to small systems or reliable/static networks, have been shown to break down when computing and network resources are scare and networks are highly dynamic.

G. Moro, S. Bergamaschi, and K. Aberer (Eds.): AP2PC 2004, LNAI 3601, pp. 201–212, 2005.

Peer-to-peer systems, such as those used for Internet file sharing (e.g. Napster or Gnutella) appear to offer a solution to these problems as they allow queries over large networks. However, these solutions have problems of central points of failure [1] or partial lookup [2]. Improvements to these approaches addressing these problems include several types of Distributed Hash Table (DHT) algorithms, including Chord [3] and Content Addressable Network (CAN) [4]. Currently, researchers are in active pursuit of ways to adapt these methods to mobile, ad hoc networks and other dynamic domains.

This paper shows how to develop a CAN for a mobile-agent system. We present a theoretical formulation that takes into account the potential for mobility of agents that manage data storage and data indices. This approach has been validated using the Extensible Mobile Agent Architecture (EMAA) from Lockheed Martin Advanced Technology Laboratories.[5] The paper presents empirical results showing how agents in an EMAA network can use an agent-based CAN to achieve the level of scalable data lookup needed for large-scale multi-agent systems applications. We present a study of the performance of the CAN on a simulated network and show that performance of the CAN implementation in the mobile agent context is in line with known performance in fixed networks. In this way, CANs can be adapted to agent systems with a performance compromise. The paper concludes by discussing a few related complex problems posed by data and index mobility in the CAN, as well as the effects of dynamic, ad hoc networks.

2 Agent-Based Data Lookup Problem

A mobile, intelligent agent operating alone is an effective tool, but a group of intelligent agents working together is potentially capable of far more, able to exploit the strengths of individual agents and network resources to accomplish their collective tasks more quickly and efficiently. In order to collaborate, agents need access to several capabilities. One of these capabilitiesis efficient data exchange and lookup.

For two agents to work together on the same task it is often necessary for one agent to tell another about the results of one subtask before another subtask can be undertaken. Various blackboard or whiteboard systems have been created to facilitate this type of collaboration, but this approach yields only a local data lookup system, forcing agents to migrate to a particular location to perform this data exchange. Besides the obvious inefficiencies of such a mechanism, the centralized nature of the whiteboard is undesirable in the face of unreliable networks.

We are interested not only in enabling two agents to collaborate, but in allowing an entire system of agents to collaborate. We can define this agent system as the set of agents $A = \{a_1, a_2, a_3, \ldots\}$, the set of network nodes $N = \{n_1, n_2, n_3, \ldots\}$, and the set of data segments they wish to share $D = \{d_1, d_2, d_3, \ldots\}$. At any given time t, each datum $d \in D$ and agent $a \in A$ are located at some network node $n \in N$. In order to maintain efficient lookup, this system must store a mapping from data and agent to index, I.

It is also important to consider the topology of the network being used. This topology as it exists at any time t is described as the set of nodes in the network, N and the set of direct connections or links, $L \subseteq N \times N$. In dynamic networks, an environment in which agents are often applied, the sets L and N change over time, as network nodes and connections are dropped. The entire agent system using a data-lookup service can now be described as the 5-tuple $\{A, N, L, D, I\}$. The index I that allows agents migrating between network nodes to locate data available on remote network nodes must be maintained in the face of these dynamic networks. Additionally, individual network nodes are often resource constrained, particularly in mobile environments where storage and processing power are at a premium. The structure and maintenance of this index is the focus of this work.

3 Background

3.1 Intelligent Agents

Intelligent agents, self-contained software entities able to utilize resources available on a network to perform particular tasks, have become a popular mechanism for collaboration between heterogeneous systems. In particular, mobile, intelligent agents are capable of traveling between systems to perform various parts of their task and to interact with existing systems. Such an agent operates by utilizing an agent framework to execute on one machine then transfer itself to another machine elsewhere on the network. It is then able to utilize whatever resources and services exist on the new machine.

These services available to the intelligent agents are essential to allowing the agent to perform its task. A service might allow access to a local database or to perform a certain type of complex calculation. Additionally, some services allow agents to coordinate with one another and collaborate to accomplish tasks, such as the whiteboard and other data lookup services mentioned previously.

One of the key advantages to the use of mobility in agents is the ability to transport data and services to the computing resources at which they are required. This advantage, however, is a difficulty to any data-indexing approach, due to the dynamic nature of the data and services being indexed. This dynamic nature must be taken into consideration in the construction of a data-lookup mechanism for an agent-based system.

Past intelligent agent applications include many data-retrieval and legacy-system integration applications. This technique has been particularly useful in providing data-retrieval to and from heterogeneous legacy systems. These types of systems commonly operate on low-bandwidth networks and with resource-constrained computing resources. As such, agents' ability to exploit available resources on a network are key to adequately performing the required tasks in these environments.[6]

Several different agent frameworks have been developed that could be used for such an application. These frameworks include the Extensible Mobile Agent Architecture (EMAA), the Cognitive Agent Architecture (Cougaar)[7], the CoABS Grid [8], Reusable Environment for Task Structured Intelligent Network Agents (RETSINA)[9], the Distributed Environment Centered Agent Framework (DECAF)[10], Aglets Workbench [11], the Decentralised Information Ecosystem Technologies Agents Platform (DIET)[12], as well as others. Each of these platforms has different strengths and weaknesses and are therefore better suited to different applications. However, these various frameworks exemplify the various capabilities of agents in general.

3.2 Distributed Hash Tables

File sharing networks such as Gnutella and Napster address much the same lookup problem faced by agents. Instead of files to exchange, however, agents are interested in the location of results, services, or other agents. Exactly what is being sought, however, is largely irrelevant, as the lookup itself is the difficult task.

Gnutella and Napster, however, do not provide the final solution to the data lookup problem. While they answer the question of decentralization, they are not massively scalable, and in fact, must trade off accuracy for efficiency. Several different approaches have been taken to create a fully scalable, decentralized data lookup system. Since such a lookup system essentially provides a Hashtable-like interface distributed over a network, such a system has been termed a Distributed Hashtable or DHT. Recent DHT approaches include Chord and Content Addressable Networks (CAN), which each have their own benefits and drawbacks.

3.3 Content Addressable Networks

This research focuses on Content Addressable Networks for their simplicity and efficiency. In a CAN, the Hashtable's key space is divided into a n-dimensional, toroidal surface. This surface is then divided into Zones, and each network node is responsible for one or more of these Zones. When data is placed into the CAN, a hash key is generated for each dimension, and this n-dimensional key defines its location within the CAN. To find the data later, the hash key can be regenerated, and the appropriate Zone can be queried to retrieve the data.

To facilitate routing of queries between network nodes, each Zone must know about neighbors in each dimension. Therefore, each zone knows of at least one neighbor in the positive direction and one in the negative direction on each dimension. Because the space does not always divide evenly, it is possible for one zone to have multiple neighbor zones in a given dimension. In this case, all of those neighbors are tracked.

When a query is made at a network node for data at a particular location, this query can simply be forwarded to the local Zone's neighbor that is nearest to the data. Each neighbor zone's coordinate range is examined, and the zone with the minimum distance from any point within its range to the query location is considered the nearest zone. The nearest neighbor Zone can then continue forwarding the request until it reaches the correct location and the data is found and forwarded to the querying node. This progression is shown in Figure 1.

In order to handle dynamic networks, a CAN must be able to handle events such as node arrival and node departure, either announced or accidental. When a new node arrives to join the CAN, an existing Zone can simply be split into two new Zones. This is achieved by transferring responsibility for half the data to the new Zone as well as reassigning neighbors for the new and old Zones, as they will have changed. When a node departs, another node within the network must take responsibility for the departing node's Zone. This is how a node can become responsible for multiple Zones. If the departing node made an unannounced disconnection, any data stored only on that node's zones may be lost, but this risk can be mitigated by redundant storage of data at multiple locations.

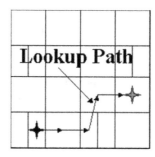

Fig. 1. A lookup message travels between neighboring zones until it reaches the zone containing the requested data

Redundant data storage is just one of many robustness and efficiency improvement techniques that can be used with CANs. It is also possible to account for temporary disconnections by allowing Zones to forward data queries through multiple neighbors, minimizing the risk that the query will not reach its target, at a bandwidth cost. To improve efficiency, it is possible to intelligently decide which Zone should be split to accommodate a new node based on its network latency to the neighbors it would inherit. Approaches such as these are mainly for future work but the potential advantages are worth noting.

4   A Mobile-Agent-Based CAN

4.1   Requirements

To allow the use of a CAN in a multi-agent system, it must be accessible to agents and servers in the system from the location of any agent in the system. Each node in the agent system must maintain a portion of the CAN index, contained in one or more CAN Zones. Each node must provide a CAN service to allow agents access to the CAN. These local services will need to communicate with each other to maintain the overall index as well as pass along data lookups. Since these services exist within the agent framework, they should ideally exhibit the same robustness and flexibility provided by the overall agent framework, including stability in the face of changing network configurations and network node failures.

4.2   Problem Formulation

An agent-based CAN implementation is much like any other CAN implementation. The CAN is made accessible to agents by publishing it as a data-lookup service. Interfaces are created to allow agents to publish as well as retrieve data through the CAN service.

Any CAN implementation requires messages to be passed between nodes in the CAN. These messages perform lookups, data registration, as well as CAN-maintenance functions, including addition of nodes to the network and zone takeover. In an agent-based CAN, instead of sending messages, communication can potentially take place by dispatching lightweight agents between nodes in the network, though this is not strictly necessary. This allows additional robustness in the network through such mechanisms as allowing an agent to select an alternate route due to node or network failures. Additionally, this allows a single lightweight agent to follow the entire path of a lookup as well as perform the final retrieval of the desired data.

Our approach consists of three components. The first of these is the DHT Server, the agent service through which agents, A, in the system interact with the CAN. One DHT Server exists on each network node, n. Each DHT Server is responsible for maintenance of one or more Zones, which are the second component of the system. A Zone z is responsible for one d-dimensional region within the DHT and maintains the locations of the data $\{D_z \subseteq D\}$. The set of these zones is then responsible for maintaining the index I. The third component is made up of the messaging agents that maintain the DHT and communicate data lookups and registrations between DHT Servers.

In theory, CANs are a highly scalable, efficient data lookup mechanism. Data does not have to be duplicated across multiple nodes in order to guarantee successful lookup. Additionally, data lookup latency is highly efficient ($O(d \mid N \mid^{1/d})$), where n is the number of nodes in the network and d is the dimension of the network. For applications where large numbers of nodes are expected, a large d can be selected in order to increase the scalability of the system at a slight per-node space cost (linear with respect to number of dimensions) to maintain a more complex network.

5   Implementation

5.1   The Extensible Mobile Agent Architecture (EMAA)

As noted earlier, an intelligent, mobile agent requires an agent framework, or run-time environment, in which to operate. For this research, the Java-based Extensible Mobile Agent Architecture (EMAA) was chosen. EMAA operates by providing a Dock, which executes mobile agents on a local machine and provides the mechanisms by which agents can migrate to remote Docks as well as interact with locally provided services.

EMAA also provides the Distributed Event Messaging System (DEMS), which allows EMAA servers and agents to perform lightweight communication across multiple Docks using messaging agents.[13] DEMS is built to resemble Java's built-in event system. Servers and agents throw DEMS events like they would any other Java events, and the built-in EMAA distribution mechanisms deliver the events to any agent or server which has registered to receive that type of event. The main advantage of using DEMS as our message passing service is the robustness provided by its agent-based delivery.

5.2   EMAA Distributed Hashtable Server

For this research, we implemented an EMAA DHT Server wthat resides on each network node. EMAA agents and servers communicate with the CAN through this DHT Server, which provides the same functionality of a standard Hashtable, specifically that of storing data and looking up data. Since this interface is an EMAA Server, any agent or server is able to interact with it as they would any local service such as a local whiteboard, oblivious to the fact that data may be stored or found remotely.

This DHT Server contains at its heart a Java-based CAN. First, a CanNode class was created which implements the basic functions required to manage the interactions

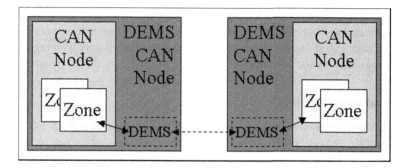

Fig. 2. The agent-based CAN is formed using DEMSCANNode, CANNode, and Zone objects on each network node

of nodes within a CAN, such as the joining of new nodes to the network. This class initially makes the assumption that it can make direct function calls to other CanNodes, as if they are executing locally. This assumption is alleviated by the extensions in the DemsCanNode, where these direct function calls are replaced by events sent through DEMS lightweight messaging agents. Each CanNode contains a Zone (or Zones), which is responsible for the portion of the index stored at that node and the IDs of neighbor zones. This structure can be seen in Figure 2.

5.3   Indirect DHT Lookup

A DHT designed to work with mobile agents must take into account that data indexed by the DHT is often carried from node to node by the mobile agents. If a standard DHT is used directly, each time an agent migrates to a different node, the DHT's index for each piece of data carried by the agent must be updated. If an agent is highly mobile, this can become an extremely large amount of data registration message traffic. One approach that could help this somewhat would be the use of the CAN structure itself to facilitate multicasting, but while this would reduce the number of messages exchanged, each node in the system would still need to hear about each agent migration. [14]

   In order to alleviate this problem, an Indirect DHT Server was created. This server incorporates two separate DHTs spread across the entire network. The first maps a data's key to an agent ID. The second maps the agent ID to an agent's current location. This requires two lookups to be performed each time data must be found. However, instead of needing to update each datum held by the agent's index in the DHT, only one entry in the agent to location mapping must be updated when the agent migrates. Because of the additional lookup time required, this Indirect DHT is only appropriate for use in systems of highly-mobile, data-heavy agents.

6   An Agent-Based DHT Server Testbed

6.1   Experimental Testbed

In order to test the agent-based CAN, we needed to set up a group of EMAA Docks in which our agents could migrate. For our tests up to 50 EMAA Docks were started on a single machine. Each Dock operated within its own Java Virtual Machine and communicated with other Docks as though each was on its own machine only connected by the network. Test Agents were created which were tasked only with registering and requesting data on fixed schedules.

6.2   Evaluation Metrics

As with most network-related services, the most important aspects of a CAN implementation are latency and bandwidth. In this case, we measure latency as the average number of hops a query must take to reach the data sought. As the current implementation of the CAN does not take into account the time latency between nodes on the network, we do not consider differences in time between node links.

Bandwidth in this case is measured as the total number of messages exchanged. This must account for both the lookup messages and any network maintenance messages created in the system, such as for adding and removing nodes from the system.

6.3   Experimental Results

Node Scalability. One of the most important factors in the usefulness of a DHT is its scalability. As such, it was important to verify that the EMAA CAN server exhibited the expected behavior of a CAN-style DHT. Specifically, the expected average lookup time should be $O(d \mid N \mid^{1/d})$ where d is the dimension of the CAN and N is the set of nodes participating in the CAN.

Since this research explores an implementation of a CAN within a set of separate EMAA docks rather than a simulation, this limited the number of nodes that could be deployed to participate in the CAN. Therefore, CANs were set up consisting of 5-50 nodes in both a 2-dimensional and 3-dimensional network.

In each condition, an agent is deployed which places 1000 random items into the DHT and then migrates to 1000 random locations and performs a lookup to retrieve one of the previously placed items.

During this process, the number of messages to look up data were recorded for each condition. It was expected that this value would follow the expected lookup-length trend of $O(d \mid N \mid^{1/d})$ for a CAN-style DHT. Additionally, the total number of messages sent was recorded for each condition. This includes all network setup messages as well as the data registration and lookup messages exchanged.

In both the 2-dimensional and 3-dimensional experiments, it was found that the actual number of lookups approximately follows the expected number of lookups of $O(d \mid N \mid^{1/d})$. There is some variance seen, however. This is believed to be due to the uneven sizes of Zones within the d-dimensional space necessary to completely cover the space as well as the random nature of the keys selected for registration to the CAN.

Additionally, as expected, larger dimensional networks require a larger number of total messages for smaller networks due to their larger setup overhead. Conversely, smaller dimensional networks require a larger number of total messages for larger networks due to their less efficient lookups.

These results can be seen in Figure 4, which shows the actual as well as expected number of lookup messages required to perform lookups in networks of 5 to 50 nodes.

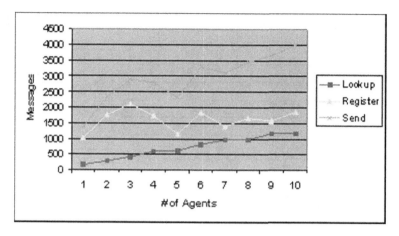

Fig. 3. As the number of agents in the system is increased, the number of lookup messages exchanged increases linearly, as expected

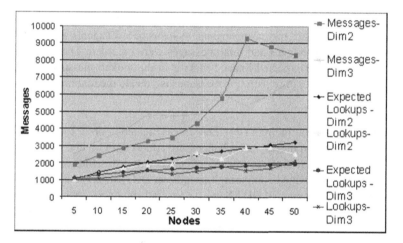

Fig. 4. In both 2-dimensional and 3-dimensional CANs, the agent-based implementation shows the expected lookup efficiency

It also includes the total number of messages used in these networks, including CAN maintenance messages.

Agent Scalability. As important as it is for the DHT to remain efficient as the network grows larger, it is equally important that the DHT maintain its performance as the number of agents accessing the DHT increases. Since agents which wish to collaborate must in the worst case communicate with each other agent in the system, in the worst case, the number of messages required for this collaboration could grow exponentially.

Fortunately, since the agents in our system are interacting only with the DHT, the number of messages required should only grow linearly as the number of agents in the system is increased. Thus, our second experiment tested the number of messages

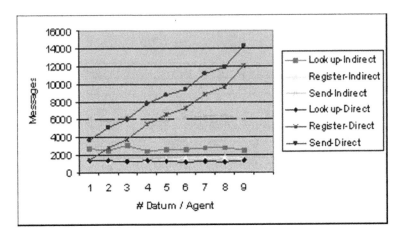

Fig. 5. The Indirect DHT Server provides more efficient registration for highly mobile, data-heavy agents, but causes lookup times to increase twofold in all cases

exchanged in systems of increasing agents. Systems of 5-40 agents were tested and the number of lookup messages and total messages were recorded. The results, shown in Figure 3 confirm the expected linear behavior in the lookup messages, though some variance in the number of registration messages is s due to the random nature of the key selection.

Indirect Lookup. The Indirect DHT Server was created in order to alleviate the large number of registration messages exchanged in systems with highly-mobile agents carrying large amounts of data. To test the effectiveness of this Indirect DHT Server, an experiment was performed comparing the Indirect DHT Server to a standard or direct DHT Server. For each type of DHT Server, a varying amount of data is assigned to a set of 10 mobile agents. Each agent migrates between nodes on the network, reregistering the location of the data it carries and performing a lookup of a single data item. Agents are dispatched containing from 1 to 9 data items each.

Data lookup messages, data registration messages, and the overall number of messages were measured for each configuration. The number of registration messages exchanged in the Direct DHT configurations was linearly related to the number of data items per agent. In the Indirect DHT configurations, the number of registration messages was constant. In both configurations, the number of lookup messages was constant but the Indirect configurations took twice as many lookup messages, since both the agent and location required lookup. These results can be seen in Figure 5.

Based on these results, both the Direct and Indirect DHT Servers seem to be useful in different situations. For systems with agents carrying large amounts of data, the Indirect Server resulted in fewer overall messages. Specifically, for agents performing a single lookup at each network node, if agents are carrying at least 3 data items, using the Indirect DHT Server results in fewer overall messages. Otherwise, it is more efficient to use the simpler Direct DHT Server.

7   Discussion and Future Work

As mentioned earlier, one of the future areas of work that should be pursued is implementation of the various robustness and efficiency mechanisms proposed for CANs and other DHTs. These would allow the system to be much more deployable in a real-world, non-laboratory environment.

Additionally, it would be valuable to perform tests of the Agent-Enabled CAN on an actual ad hoc tactical network, such as that available by the Secure Wireless Agent Testbed (SWAT) project which distributes an agent-based system across multiple iPAQ platforms on a wireless network.[15] This would validate both the CAN implementation itself and the robustness techniques.

In ad hoc networks, mobile agent systems can provide a level of robustness in the face of network realignment and node failure. To this end, services can be relocated or redistributed across network nodes based on available resources. In such an environment, CAN Zones could be redistributed across different sets of network nodes. To provide this mobile index, it will be necessary to develop an algorithm to determine how and when to redistribute portions of the CAN Zones across nodes.

If the agents, data, and index are simultaneously mobile, many maintenance messages will be required between CAN Servers. Additional work should be done to minimize these maintenance messages. Such a reduction could possibly be provided by combining messages or by eliminating redundant messages. Additionally, the inherent multicast capabilities of a CAN could be exploited to more efficiently deliver additional maintenance messages.

8   Conclusions

This research has developed an adaptation of Content Addressable Network (CAN) for mobile agents and implemented the approach within the Extensible Mobile Agent Architecture (EMAA). The CAN allows an efficient, scalable data lookup mechanism for mobile agents, specifically supporting the kind of distributed data storage needs for mobile and ad hoc networking environments. Empirical validation shows that this implementation provides the expected scalability of $O(d \mid N \mid^{1/d})$ for both two and three dimensional CANs. An Indirect DHT implementation was shown to provide increased efficiency for systems of highly-mobile agents.

Additionally, several future enhancements to the implementation have been identified. These improvements will provide a CAN which is truly capable of handling the specific characteristics of multi-agent system, particularly that of those operating in dynamic, unreliable environments.

References

1. S. Saroiu, P. Gummadi, and S. Gribble. A measurement study of peer-to-peer file sharing systems. In *Proceedings of the Multimedia Computing and Networking (MMCN)*, San Jose, CA, January 2002.
2. J. Ritter. Why gnutella can't scale. no, really. Available from http://www.darkridge.com/ jpr5/doc/gnutella.html.

3. I. Stoica, R. Morris, D. Karger, M. Kaashoek, and H. Balakrishnan. Chord: A scalable peer-to-peer lookup service for internet applications. In *Proceedings of ACM SIGCOMM*.

4. S. Ratnasamy, P. Francis, M. Handley, R. Karp, and S. Shenker. A scalable content-addressable network. In *Proceedings of the 2001 conference on applications, technologies, architectures, and protocols for computer communications*, August 2001.

5. R. Lentini, G. Rao, J. Thies, and J. Kay. Emaa: An extendable mobile agent architecture. In *AAAI Workshop - Software Tools for Dev Agents*, July 1998.

6. S. McGrath, D. Chacn, and K. Whitebread. Intelligent mobile agents in the military domain. In *Fourth International Conference on Autonomous Agents 2000*, June 2000.

7. BBN Technologies. Cougaar architecture document. Technical report, BBN Technologies, 2003.

8. M. Kahn and C. Cicalese. The coabs grid. In *Innovative Concepts for Agent-Based Systems: First International Workshop on Radical Agent Concepts, WRAC-2002*, volume LNCS 2564 of Lecture Notes in Computer Science, 2002.

9. K. Sycara, M. Paolucci, M. vanVelsen, and J. Giampapa. The retsina mas infrastructure. *Journal of Autonomous Agents and Multi-agent Systems (JAAMAS)*, pages 29–48, 2003.

10. J. Graham, K. Decker, and M. Mersic. Decaf - a flexible multi-agent system architecture. In *Autonomous Agents and Multi-Agent Systems*, pages 7–27, 2003.

11. D. Lange, M. Oshima, G. Karjoth, and K. Kosaka. Aglets: Programming mobile agents in java. In *International Conference on Worldwide Computing and its Applications (WWCA '97)*, volume LNCS 1274 of Lecture Notes in Computer Science, 1997.

12. P. Marrow, M. Kobarakis, R. van Lengen, F. Valverde-Albacete, E. Bonsma, J. Cid-Suerio, A. Figueriras-Vidal, A. Gallardo-Antoln, C. Hollie, T. Koutris, H. Molina-Bulla, A. Navia-Vzquez, P. Raftopoulou, N. Skarmeas, C. Tryfonopoulos, F. Wang, and C. Xiruhaki. Agents in decentralised information ecosystems: the diet approach. In *Proceedings of the Artificial Intelligence and Simulated Behaviour Conference 2001 (AISB '01), Symposium on Information Agents for Electronic Commerce*, pages 109–117, 2001.

13. J. McCormick, D. Chacn, S. McGrath, and C. Stoneking. A distributed event messaging system for mobile agent communication. Technical Report TR-01-02, Lockheed Martin Advanced Technology Labs, March 2000.

14. S. Ratnasamy, M. Handley, R. Karp, and S. Shenker. Application-level multicast using content-addressable networks. In *Lecture Notes in Computer Science*, volume 2233, pages 14–25, 2001.

15. G. Anderson, D. Artz, V. Cicirello, M. Kam, N. Morizio, A. Mroczkowski, M. Peysakhov, W. Regli, and E. Sultanik. Secure mobile agents on ad hoc wireless networks. In *The Fifteenth Innovative Applications of Artificial Intelligence Conference*, Aculpulco, Mexico, August 2003.

Intelligent Agent Enabled Genetic Ant Algorithm for P2P Resource Discovery

Prithviraj(Raj) Dasgupta

Department of Computer Science,
University of Nebraska, Omaha, NE 68182
pdasgupta@mail.unomaha.edu
Phone: (402) 554 4966 Fax: (402) 554 3284

Abstract. Rapid resource discovery in P2P networks is a challenging problem because users search for different resources at different times, and, nodes and their resources can vary dynamically as nodes join and leave the network. Traditional resource discovery techniques such as flooding generate enormous amounts of traffic, while improved P2P resource discovery mechanisms such as distributed hash tables(DHT) introduce additional overhead for maintaining content hashes on different nodes. In contrast, self-adaptive systems such as ant algorithms provide a suitable paradigm for controlled dissemination of P2P query messages. In this paper, we describe an evolutionary ant algorithm for rapidly discovering resources in a P2P network.

Keywords: Peer-to-peer systems, software agents, ant algorithm, adaptive systems, genetic algorithm.

1   Introduction

Over the past few years, peer-to-peer(P2P) systems have emerged as an attractive communication paradigm between users in a networked environment. In a P2P network, every node behaves as a peer with similar functional capabilities. This makes P2P networks suitable for connecting large numbers of users in a distributed manner without worrying about scalability and centralized control issues. One of the main challenges in P2P networks is to enable rapid and efficient discovery of resources present on the different nodes of the network. In this paper we describe an evolutionary algorithm inspired by the foraging behavior in insect colonies such as ants to enable rapid exploration of the search space.

In most commerical P2P systems[6], resource discovery is implemented by flooding a resource query across nodes of the network. Flooding generates considerable traffic and ensues network congestion. Improved P2P resource discovery algorithms employ super-peer nodes[14] and dynamic hash tables(DHT)[12,8] to strategically place resources on nodes to enable rapid lookup. These techniques address P2P resource discovery as a resource management problem. In contrast, the genetic ant algorithm described in this paper uses information obtained from previous resource queries to improve future searches. Ant algorithms have been

G. Moro, S. Bergamaschi, and K. Aberer (Eds.): AP2PC 2004, LNAI 3601, pp. 213–220, 2005.

applied to several hard problems[2,4] such as dynamic programming, the travelling salesman problem and routing in telecommunication networks[5]. Resource discovery in P2P networks is different from each of these problems because the node on which the resource will be discovered is not known apriori and the topology of the P2P network changes dynamically as nodes join and leave. Anthill[1] employs ant-based algorithms for load balancing in a P2P network and ants travel across the network to update routing tables at each node. In contrast, our algorithm uses different types of pheromone and ants with different behavior to make P2P resource discovery more efficient. Genetic ant algorithms have been researched in [3,13,9] to evolve parameters of the algorithm itself. In this paper, we use an evolutionary ant algorithm to evolve new routes in a P2P network.

2   Ant Algorithm for Peer-to-Peer Resource Discovery

The primary objective of a node in a P2P network is to search and acquire resources available on other nodes in the network, and, simultaneously allow other nodes to access resources present on the node itself. The traditional P2P resource discovery protocol consists of a *query* message that is forwarded to successive nodes in a breadth-first manner across the network until the resource is discovered or the lifetime of the message expires. If the resource is found on a node a *queryHit* message is sent back from the node possessing the resource to the node that originated the query. The requesting node and the providing node then decide on the download protocol for the resource.

The P2P resource discovery protocol can be made more efficient if the uninformed search in the traditional protocol can be provided with a heuristic-based informed search. Our informed search algorithm for P2P resource discovery is inspired by the foraging activity used by social insects such as ants[2] to locate food sources. In several ant species, foraging ants searching for food leave behind a pheromone trail along the path from their nest to the food source. Ants searching for the food source later on use the trail as a positive reinforcement to lead themselves to the food source.

The traditional ant algorithm uses pheromone as a positive reinforcement on nodes visited by an ant which lead to a solution. We envisage that P2P search can be made more efficient if an *anti-pheromone* is used to provide negative reinforcement to an ant at a node that does not lead to a solution. However, nodes that are not useful for locating one resource, might lead to or contain the solution for another resource. To enable exploration of nodes marked with anti-pheromone, we use another type of ant called *explorer* ants that are are neutral to nodes with pheromone but get attracted to nodes with anti-phermone.

Based on these functionalities we have used the following types of ants in our algorithm: **(1) Forward Foraging Ants** visit nodes searching for a resource and deposit pheromone at each node they visit. From each node, a foraging forward ant prefers to go to neighbor nodes that have higher amounts of phermone and lesser amounts of anti-pheromone. **(2) Forward Explorer Ants** visit nodes searching for a resource and deposit anti-pheromone at each node they visit.

From each node, a forward explorer ant prefers to go to neighbor nodes that have higher amounts of anti-phermone. **(3) Backward Ants** Both types of forward ants become a backward ant when either they discover the resource they are searching for on a node, or, they reach the search boundary without discovering the resource. Backward ants trace the route taken by their corresponding forward ant in the reverse direction. A backward ant deposits pheromone at each node it visits if the resource that the corresponding forward ant was looking for was found, and, deposits anti-pheromone if the resource was not found.

3   P2P System Model

Our model of the P2P system comprises a connected network of N nodes. Nodes join and leave the network at random. Each node maintains a forwarding table containing the addresses of its neighbor nodes determined using the P2P node discovery protocol. Each address in the forwarding table is associated with a normalized weight that represents the pheromone associated with that node. The weight of a node in the forwarding table gets updated when an ant selects it to move to it. Pheromone increases the weight while anti-pheromone decreases it. When a user at a node enters a query to search for a resource in the P2P network, an ant gets created on the origin node for the query. The ant visits different nodes of the network searching for the resource using the ant algorithm described below.

3.1   Ant Algorithm

Forward Foraging Ant. The algorithm used by a forward foraging ant at a node n to select a neighobor node i and update the weight associated with node i uses the following parameters:

a_n	Number of nodes in the forwarding table of node n
$w_{i,n}^t$	Normalized weight associated with neighbor node i of node n at time t
τ_n	Amount of pheromone deposited on node n
τ_0	Amount of pheromone deposited by ant at source node of the search
χ_n	Amount of anti-pheromone deposited on node n
χ_0	Amount of anti-pheromone deposited by ant on node at which search boundary was reached after resource was not located
$h_{s,n}$	Number of hops made by an ant to reach from the node s on which it started its journey to the current node n

The update rules for the pheromone at node n are the following:

$$\tau_n = \frac{\tau_0}{[h_{s,n}]^\alpha}$$
$$w_{i,n}^t = w_{i,n}^{t-1} + \tau_n(1 - w_{i,n}^{t-1}) \tag{1}$$

where s is the origin node that initiated the search query.

$$w_{i,n}^t = \frac{w_{i,n}^t}{\sum_{i=1}^{i=a_n} w_{i,n}^t} \tag{2}$$

The factor α is determined expermentally and it controls the decrease in the amount of pheromone deposited as the ant moves further away from its origin. The second term on the r.h.s of Equation 1 ensures that the amount of pheromone deposited on a node is proportional to its current weight. This prevents excessive phermone (or anti-pheromone) being deposited on a node whose weight is very high(or low). Equation 2 ensures that the weights of nodes in the forwarding table remain normalized after the weight of a node is updated by an ant.

Forward Explorer Ant. A forward explorer ant works in a manner similar to a forward foraging ant except that it uses the inverse probability $(1 - w_{i,n}^t)$ to select a node i from the forwarding table of its current node n. This ensures that the probability of selection of a node by an explorer ant is proportional to the amount of anti-pheromone deposited on it. A forward explorer ant updates the anti-pheromone at each visited node according to the following equations:

$$\chi_n = \frac{\chi_0}{[h_{s,n}]^\alpha}$$
$$w_{i,n}^t = w_{i,n}^{t-1} - \chi_n(1 - w_{i,n}^{t-1}) \tag{3}$$

where s is the origin node for the explorer ant.

Backward Ant. When a forward ant locates a resource or reaches its search boundary without locating the resource, it becomes a backward ant. If the resource was located by the forward ant, the backward ant rewards each node along the reverse route with pheromone using Equation 1. Otherwise, if the search boundary was reached without locating the resource, the backward ant deposits anti-pheromone on each node it visits using Equation 3 to indicate that the node did not lead to a succesful resource discovery. For the backward ant, the node s represents the the node on which the resource was found (in Equation 1) or the node on which the search boundary was reached without locating the resource (in Equation 3).

4   Genetic Ant Algorithm

In the ant algorithm described in section 3.1 each ant-type is associated with a specific pheromone-type and the pheromone(anti-pheromone) along a particular trail keeps on increasing as more foraging(explorer) ants follow that trail. Eventually, the paths in the network get partitioned into trails that are predominant either in pheromone or in anti-pheromone, with the traditional ant algorithm (one type of ant, one type of pheromone) running within each partition. This problem can be addressed if two ants, each of a different type, periodically exchange the routes they have taken, even partially, with each other. This would

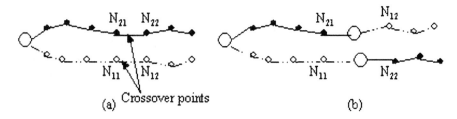

(a) Crossover points

(b)

Fig. 1. Single crossover operator used when the routes represented by the two chromosomes do not have any common nodes. (a) The routes(chromosomes) before crossover, and, (b) The origin node, O, is introduced at the crossover point of the routes(chromosomes) during crossover.

re-balance the amounts of pheromone and anti-pheromone along the trails in the P2P network and prevent a network partition based on pheromone-type. Trail exchange between ants also prevents ants from following routes that have become outdated due to the dynamic joining and leaving of nodes.

Genetic algorithms(GAs)[10] provide a suitable mechanism for implementing exchange of trails between ants. A GA enables a problem to rapidly converge to an improved solution using an evolutionary mechanism. We have adapted the traditional GA to evolve the routes taken by ants. We employ the ant algorithm described in Section 3.1, to initially discover routes to resources in the P2P network. When a certain number of search queries have originated from a particular node, a GA is run on the node using the the routes traversed by the ants created for each search query originating from the node. The different attributes used the GA are described below:

- **Fitness Function.** The fitness function F for a route taken by an ant is given by the following equation:

$$F = 1 - \frac{\text{number of hops to locate resource}}{maxHops}, \text{if the ant was successful}$$
$$F = \gamma, \text{if the ant was unsuccessful to locate the resource}$$

  The parameter γ represents the probability of recombining the children chromosomes in the next generation.
- **Chromosome Representation.** For enabling the GA, a chromosome of an ant is represented as the route (sequence of nodes) visited by the ant.
- **Crossover Operator.** The routes(chromosomes) traversed by the ants have the following characteristics: a) the routes(chromosomes) might be of different lengths b) the routes represented by the chromosomes may or may not have common nodes between them. We have identified the following two scenarios to address these issues:
  Scenario 1: The routes represented by the two chromosomes participating in the reproduction do not have any common nodes. As shown in Figure 1, we select the crossover point randomly in each of the two parent chromosomes and use single point crossover. The origin node O is introduced at the

crossover point after performing the crossover. This ensures that new entries do not need to be introduced within the forwarding tables on the nodes at the crossover points of the two participating chromosomes. *Scenario 2: The routes represented by the two chromosomes participating in the reproduction have one or more common nodes.* We use an n-point crossover operator. A crossover point is determined as a node that is common between the two chromosomes participating in the crossover. Since there can be more than one pair of nodes common between two chromosomes, multiple crossover points can exist. Therefore, we use the n-point crossover operator that exchanges the chromosomes of the parents between alternate pairs of crossover points.

– **Recombination.** We reintroduce the evolved children ants obtained through crossover into the parent population. The fitness of each child is determined as the average value of the fitness of each parent.

5   Experimental Results

The P2P network used for our simulation contains $N = 100$ nodes. The number of neighbors of each node is generated from a normal distribution with mean $N/20$ and standard deviation 1.5. The neighbor nodes for every node are then selected randomly and the connections between nodes are set up by initializing the forwarding table inside each node. Resources are then simulated on each node by adding a string identifier corresponding to the name of the resource inside the resource table of a node. A resource is added on a node with a probability ρ that denotes the availablity of the resource in the P2P network. Nodes are also added and removed dynamically as the algorithm runs to simulate the joining and leaving of nodes in a P2P network.

Figure 2(a) shows the percentage of successful search queries in the network when we vary the probability p with which a forward ant decides to forage or to explore when it is created. $p = 1.0$ means that the ant is created a forward foraging ant while $p = 0$ means that the ant is created as a forward explorer ant. When the network is initially set up there is no pheromone on any node. Therefore, at the beginning of our simulation we created all ants as foraging ants($p = 1.0$) and gradually decreased p until all ants were created as explorer ants ($p = 0$). Other parameters used for this experiment were $\rho = 0.15$,(resource availability on nodes) $\alpha = 4.0$ (pheromone decay rate with distance from ant's origin) and a *searchLimit* of 10 hops. The results of our experiments illustrate that the number of successful search queries increase when the value of p is in the range of 0.4 to 0.8 with a mean of around $p = 0.72$. The reason for this can be attributed to the fact that with $\rho = 0.15$, resources are very likely to be located before the *searchLimit* of 10 hops(among 100 nodes) is reached. Therefore, nodes that have been marked with pheromone by foraging ants right after set up are likely to lead to sucessful location of resources for subsequent searches.

Figure 2(b) shows the effect of using a genetic ant algorithm for P2P resource discovery for different values of γ, that denotes the probability of recom-

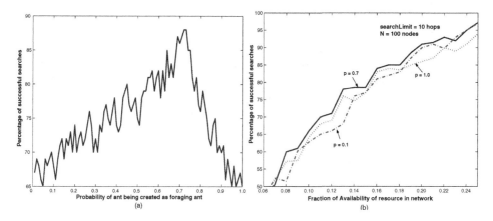

Fig. 2. (a)Percentage of successful searches in a P2P network for different probabilities of an ant being created as a foraging ant. (b)Percentage of successful searches vs. fraction of availability of resource in the P2P network (ρ) for different values ($p = \gamma$) of the probability of recombining the children chromosomes in the next generation.

bining the children chromosomes in the next generation, as the availability of the resource in the network increases. With $\gamma = 0.1$, very few unsuccessful ants are selected by the fitness function. The ant algorithm then employs only successful ants along established trails. Consequently, the ant algorithm does not perform as efficiently because ants continue to visit nodes they have already visited before. When $\gamma = 0.7$ the number of unsuccessful ants being selected by the genetic algorithm increases. As shown in Figure 2(b), the percentage of successful searches improves with $\gamma = 0.7$. With trails from unsuccessful ants being crossed-over with successful trails, ants visit more nodes on which resources have not yet been discovered to enable rapid resource search. Moreover, foraging ants depositing pheromone also visit unsuccessful trails from the previous generation that had anti-pheromone along them and vice-versa. This rebalances the amounts of the two pheromones across the network to improve the performance of the search. When γ is further increased to 1.0, unsuccessful ants and trails with anti-pheromone are selected aggresively. However, this results in ants only following previously unsuccessful trails that have anti-pheromone. Consequently, the percentage of successful searches reduces and reverts to a value similar to the performance obtained with lower values of γ. Figure 2(b) illustrates that a moderate value of γ in the range of 0.5 to 0.7 is most effective in evolving the routes in the genetic ant algorithm to enable rapid resource discovery.

6   Conclusion and Future Work

In this paper, we have described an informed search mechanism using a genetic ant algorithm for P2P resource discovery. The genetic ant algorithm is employed as a perturbation mechanism to redistribute different types of pheromone along the different routes in the network and prevents a partition of the network based on pheromone type. The efficiency of the genetic ant algorithm depends on the rate at which nodes, along with the resources within them, join and leave the network. Our simulations show that the genetic ant algorithm performs better than the traditional ant algorithm mainly when resources are scarce. In the future, we propose to investigate a cooperative multi-agent framework that allows ants from different nodes to exchange trail information with each other through a gossip mechanism to locate resources rapidly. We envisage that ant algorithms implemented through software agents provide a useful direction for further exploring challenges and issues of P2P networks for future research.

References

1. Babaoglu, O., Meling, H., and Montresor, A.: Anthill: A framework for the development of agent-based peer-to-peer systems. In: Proceedings of the 22nd International Conference on Distributed Computing Systems(ICDCS), 2002, pp. 15-22.
2. Bonabeau, E., Dorigo, M., and Theraulaz, G.,: Swarm Intelligence: From Natural to Artificial Systems. Oxford University Press, 1999.
3. Botee, H., and Bonabeau, E.,: Evolving Ant Colony Optimization. In: Journal of Advanced Complex Systems, vol 1., 1998, pp. 149-159.
4. Dasgupta, P.,: 'Improving Peer-to-Peer Resource Discovery Using Mobile Agent Based Referrals. In: Proceedings of the 2nd Workshop on Agent Enabled P2P Computing, Australia, July 2003, pp. 41-54.
5. Di Caro G.,and Dorigo, M.,: AntNet: Distributed Stigmergetic Control for Communications Networks. In: Journal of Artificial Intelligence Research, vol. 9, 1998, pp. 317-365.
6. Gnutella, URL http://www.gnutella.com
7. Kazaa, URL http://www.kazaa.com
8. Kubiatowicz, J., et al.: OceanStore: An Architecture for Global-Scale Persistent Storage. In: Proceedings of the ACM ASPLOS, 2000, pp. 190-201.
9. Monmarche, N., Ramat, E., Desbarats, L., and Venturini, G.,: Probabilistic Search with Genetic Algorithms and Ant Colonies. In: Proceedings of the Optimization by Building and Using Probabilistic Models Workshop, Genetic and Evolutionary Computation Conference, 2000, pp. 209-211.
10. Mitchell, M.,: An Introduction to Genetic Algorithms. MIT Press, 1996.
11. SETI URL http:// setiathome.ssl.berkeley.edu
12. Stoica, I., Morris, R., Karger, D., Kaashoek, F., and Balakrishnan, H.,: Chord: A peer-to-peer lookup service for internet applications. In: Proceedings of the ACM SIGCOMM Conference, 2001, pp. 149-160.
13. White, T., Pagurek, B., and Oppacher, F.,: ASGA: Improving the Ant System by Integration with Genetic Algorithms. In: Proceeding of the 3rd Genetic Programming Conference, July 1998, pp. 610-617.
14. Yang B., and Garcia-Molina, H.,: Designing a super-peer network. In: Proceedings of the 19th International Conference on Data Engineering (ICDE), March 2003.

Photo Agent:
An Agent-Based P2P Sharing System

Jane Yung-jen Hsu, Jih-Yin Chen, Ting-Shuang Huang,
Chih-He Chiang, and Chun-Wei Hsieh

Department of Computer Science & Information Engineering,
Institute of Networking and Multimedia,
National Taiwan University, Taipei, Taiwan
yjhsu@csie.ntu.edu.tw
{b89014, b89009, b89015, b89027}@csie.ntu.edu.tw

Abstract. With the proliferation of content creation devices, sharing digital contents has become an increasingly common task in our daily lives. This research proposes the "Photo Agent" that helps users manage and share digital photos without explicit file manipulation and data communication. The agents shares photos autonomously and pro-actively, so users can simply specify *which* pictures to share with *whom*, rather than *how* the pictures are actually distributed and searched. The prototype photo agent utilizes peer-to-peer networking to support efficient content sharing in a distributed environment.

1 Introduction

There has been a growing demand for easy ways to share the large amount of digitally captured contents with family and friends. InfoTrends Research Group forecasted worldwide digital camera sales to reach 53 million units sold in 2004, and to continue growing at a 15% annual rate to reach 82 million units sold in 2008[1]. While commercial or open source software packages, i.e. ACDSee[2], Paint Shop Photo Album[3], and Gallery[4] etc., help people organize digital photos on their personal computers or the web, such tools often require too much direct manipulation, thereby rendering photo sharing a chore. Some important challenges for a better photo sharing system include:

- efficient storage of a large number of images captured over time;
- efficient retrieval of specific images from the collection;
- effective content sharing mechanism given differences in user device, network bandwidth and availability;
- guaranteed delivery of contents; and
- intelligent search of contents from shared sources.

The proposed *Photo Agent* employs a human-centric approach to managing and sharing digital photos over a peer-to-peer (P2P) network. Photo Agents share photos autonomously and pro-actively, so users simply specify *which* pictures to share with *whom*, rather than *how* the pictures are actually distributed

G. Moro, S. Bergamaschi, and K. Aberer (Eds.): AP2PC 2004, LNAI 3601, pp. 221–228, 2005.

and searched. The intuitive agent-based interface enables users to be freed from file management and networking details.

In what follows, we start by presenting the overall system design architecture We then define access rights and agent communication protocol. Sections 4/5 describe details of the sharing/searching algorithms respectively. The prototype system implementation is outlined in Section 6, followed by the conclusion.

2   System Architecture

The photo agents operate on a distributed P2P network without a central server. Each node, called a *peer*, is responsible to help relay contents from the other nodes. Contents are collected and re-distributed to other peers through end-system multicast[5]. The cooperation of peers helps reduce the load on the node that sends content to a large group of friends. An *agent-based peer*[6] is an autonomous node that perceives the current network environment and makes the best decision to facilitate photo transmission and redirection.

As is shown in Figure 1, each node consists of four major components: management agent, sharing agent, searching agent, and the photo database.

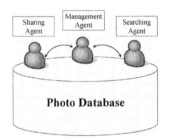

Fig. 1. Architecture of the photo agent

The *management agent* manages indexing, storage, and retrieval of digital images. It also delegates sharing and searching tasks from the user to the sharing and searching agents. The user interface provides functions for importing, sorting, and annotating photos. Duplicates photos are removed to optimize disk utilization. The *sharing agent* is in charge of distributing any set of photos to the target group as specified by the user. The cooperation of sharing agents on the network forms a sharing tree to ensure an efficient and reliable sharing process[7]. The *searching agent* helps a user to locate specific contents residing in the photo database of another peer. Depending on the searching algorithm and the access rights of the target content, multiple searching agents cooperate on the P2P network to return a group of useful results. Agents engage in active communication in carrying out the sharing and searching tasks.

3   Control and Communication

This section defines the access control and agent communication protocol in the proposed photo agent system.

In a P2P sharing system, it is necessary for users to protect their contents with access control[8]. Table 1 shows the definition of access rights in terms of four attributes[9]. The owner has complete access and can modify any attribute of his own photos. The value of *Share lifetime* decreases by one every time the photo is shared to another user. When the *Share lifetime* becomes 0, the *Search level* is set to *nobody* automatically. To preserve privacy, the *Search level* can only be adjusted lower except by the photo owner. Access right specifications should remain with a given photo, even when it is exported or shared.

Table 1. Attributes of access rights for a digital photo

Attribute	Attribute definition	Possible value
Photo owner	The user who imports the photo	The owner ID
Export right	The right to export this photo	Yes/No
Share lifetime	The number of nodes this photo can be shared	0/positive integer/∞
Search level	The level that a photo can be searched	all/friend/nobody

In sharing and searching photos, a photo agent needs to exchange status and preference information with the other agents. The basic message types include *share, request, info, report, search,* and *redirect*. The *info* message is used to communicate the capabilities and preferences of the given node. Figure 2 illustrates the sample communication transcripts of relaying reduced photos to be shared and search requests.

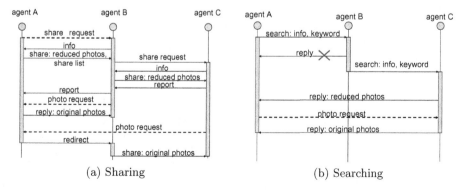

(a) Sharing   (b) Searching

Fig. 2. Agent communication

4   Photo Sharing

P2P file sharing systems, ranging from fully distributed models like Gnutella[10] to centralized models such as Napster, have gained tremendous popularity. Sharing personal experiences, in the form of images or videos, is a promising application area of P2P sharing technology. A good photo sharing system[11] needs *to distribute the contents efficiently* and *to ensure successful delivery of shared contents.* Instead of a central server that relays files to their destinations, content delivery is distributed in a P2P network. Each node is responsible for relaying photos to its peers. Figure 3(a) illustrates a sharing structure for distributing content from the *source node* A, to the *relay nodes* B and C, and then to all remaining *target nodes* D through H. The sharing agent at each node can autonomously decide the appropriate relay nodes from its sharing list[12]. To

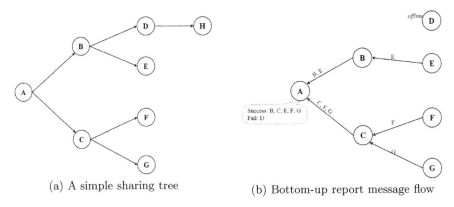

(a) A simple sharing tree   (b) Bottom-up report message flow

Fig. 3. P2P Photo Sharing

reduce bandwidth requirement, reduced images of the shared photos are propagated through the sharing tree structure. The reduced photos act as the "tickets" for requesting the original photos from the owners.

   The sharing agent on a relay node is in charge of relaying the photos to its assigned subset of target nodes. Each relay node is expected to report the successful (or failed) deliveries back to its parent node. Figure 3(b) illustrates the bottom-up reporting result if node D remains off-line. The Photo Agent system needs to ensure that each file is sent to its destination, even if the target peer is currently off-line. When node B gets no report from node D after a period of time, it reports back to the parent node A that D and H has not received the photo. The sharing agents on both nodes A and B will attempt to *catch* D and H when they get back online at a later time. Without a central server, the relay nodes have to record the unfinished sharing jobs. If a delivery is not possible when the nodes are never online at the same time, alternative delivery channels, e.g. email, may be used.

4.1 Construction of Sharing Tree

In general, the number of relay nodes R should increase with the node's bandwidth B and the number of target nodes T, i.e. $R = f(B, T)$. The function f needs to

- *Bound the file transfer time of the source node to t,* and
- *Compact the height of the sharing tree to h.*

Let n be the number of photos to be transferred and a be the average size of a typical photo image, we have $\frac{n \times a}{B} \times R \leq t$. As a result, the upper bound of R is

$$R \leq \frac{B \times t}{n \times a}. \tag{1}$$

The height of the sharing tree should be under h, that is, $1 + \lfloor \log \lceil \frac{T}{R} \rceil \rfloor \leq h$. The lower bound of R is found to be

$$R \geq \frac{T}{2^{h-1}}. \tag{2}$$

To select the best relay nodes, each target node is assigned a score s based on its current end-to-end connection quality with the source node, denoted as q, and its past recorded successes p. That is,

$$s = \alpha \cdot q + \beta \cdot p \tag{3}$$

where α and β specify the relative weights of bandwidth vs. past record.

Suppose that information about bandwidth and connection quality are aggregated in global lookup tables over time. Algorithm 1 presents the process of constructing the sharing tree. Let *sharingList* denote the list of target nodes, and *relayList* denote the list of relay nodes.

Algorithm 1. Sharing_with_global_info
Given global lookup tables Q, P and B, construct the sharing tree.

Require: *sharingList*: A list of nodes for tree construction
1: *relayList* ← source node
2: **for** $i \in relayList$ **do**
3:   **for** $j \in sharingList$ **do**
4:     $S(j) \leftarrow \alpha \cdot Q(i, j) + \beta \cdot P(j)$
5:   **end for**
6:   $R \leftarrow f(B(i), |sharingList|)$
7:   $Child(i) \leftarrow$ highest scoring R nodes from *sharingList*
8:   $relayList \leftarrow relayList \setminus \{i\} \cup Child(i)$
9:   $sharingList \leftarrow sharingList \setminus Child(i)$
10: **end for**

Algorithm 1 constructs the sharing tree by recursively selecting the R top-scoring nodes from *sharingList* to serve as the relay nodes. The construction process terminates when *sharingList* is exhausted, and the entire sharing tree is created in one shot.

When the global information about bandwidth, connection quality, and past record is not available, or when the network properties change frequently, it is not a good idea to construct a static sharing tree as described above. Instead of aggregating information in global tables, each node owns a local table that contains the information of a limited number of nodes. Given such incomplete information, the sharing tree is constructed as a result of cooperation by all nodes involved. Each node probes the others to retrieve and update the necessary information. In contrast to Algorithm 1, which builds the complete sharing tree at the root node, each node contributes in Algorithm 2 by constructing a part of the sharing tree with *k-level lookahead*. Parameter *sharingTree* is the tree structure under construction, while parameter *sharingList* is a list of target nodes to be assigned to the *sharingTree*.

First, the current node decides the appropriate number of relay nodes and the levels of sharing tree it plans to build. Algorithm 2 then proceeds with constructing the sharing tree and all related nodes. Meanwhile, information in the local lookup table is updated through cooperation. Finally, the updated sharing tree and sharing list are sent to each relay node to continue the construction of the sharing tree.

Algorithm 2. Sharing_without_global_info
Construct the sharing tree with k-level lookup.

Require: *sharingTree, sharingList*
1: Decide r, the number of relay nodes from the current node.
2: Decide k, the levels of sharing to be constructed from the current node.
3: **for** $i = 1$ to k **do**
4:    **for** $y \in$ next i-level relay nodes **do**
5:       Select the best r nodes as the relay nodes.
6:       Construct the next level of sharing trees from node y.
7:    **end for**
8: **end for**
9: Update *sharingTree* and *sharingList*.
10: Send the updated lists to the relay nodes at the next level.

The sharing agents distribute reduced images of the pictures to be shared using the sharing tree. The target user is then notified about the newly shared pictures. Requests for the original photos are only made when the user decides to accept the selected content.

5   Photo Searching

An alternative mode of sharing is to allow users to *search* for specific photos from the collections maintained by their peers. Given a search request, the *searching agent* broadcasts the request to its peers[12]. Each request message should include the following fields.

RootID: ID of the agent making the original request.
Deadline: The maximum time for request propagation.
Query: Specification of the target images.
Relay: Whether the request is a relay search.

To eliminate redundant search requests, path information is recorded so that a request will not be sent to the same node twice. When a match is found, the corresponding reduced image is sent back to the requesting agent along with the original query and agent ID. The user can then select from the matching photos to request originals from their owners. Figure 3 illustrates the searching algorithm. *Line 2* queries the local collection using `local_search`. *Lines 3-7*

Algorithm 3. Searching
Searching Algorithm.

Require: *RootID*, *Deadline*, and *Query*
1: **if** *Deadline* > 0 **then**
2:     *result* ← local_search(*Query*)
3:     **for** $z \in result$ **do**
4:         **if** access_right(z) **then**
5:             send_photo(*RootID*, z)
6:         **end if**
7:     **end for**
8:     Forward the request with a reduced deadline to everyone on the friend's list.
9: **end if**

---

check the access rights for legal distribution. *Line 8* forwards the request to all peers for more matching results.

6   Implementation and Concluding Remarks

The Photo Agent is designed to provide a simple interface that helps users manage photos conveniently, share photos efficiently, and protect photos safely. A prototype system has been developed to support content sharing in a serverless decentralized peer-to-peer network. Figure 4 shows the browser interface implemented in PHP with MySQL for data management. P2P sharing/searching is built on top of JXTA[13]. Experiments of the prototype photo agents were carried out using desktop and laptop PCs running Windows XP. Initial results show that agent-based P2P sharing is a light-weight and efficient approach to content management and distribution on the Internet.

As the population of digital camera owners grows rapidly, photo agents will become an indispensable tool for ad hoc photo sharing. Further experiments and analysis are necessary to improve the sharing and searching algorithms to

 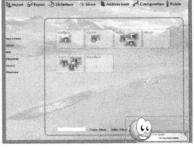

(a) Sharing selected photos   (b) Notification

Fig. 4. Photo Agent Interface

ensure fast and reliable content delivery. The proposed agent-based P2P sharing mechanism can be readily extended to other kinds of digital content in the near future.

This research is sponsored in part by the ROC National Science Council grant NSC-93-2213-E-002-087.

References

1. InfoTrends Research Group: Worldwide consumer digital cemera forcast (2003)
2. ACD Systems: ACDSee 6.0 (2004) http://www.acdsystems.com/.
3. Jasc Software: Paint shop photo album 4 (2004) http://www.jasc.com/products/photoalbum/.
4. SourceForge.net: Gallery 1.4.2 (2004)
5. Chu, Y.H., Rao, S.G., Zhang, H.: A case for end system multicast. In: ACM Sigmetrics. (2000) 1–12
6. Russell, S., Norvig, P.: Artificial Intelligence: A Modern Approach. Second edn. Prentince Hall (2002)
7. Nagi, K., Elghandour, I., Konig-Ries, B.: Mobile agents for locating documents in Ad Hoc networks. In: Proceedings of the Workshop on Agents and Peer-to-Peer Computing at AAMAS 2003. (2003)
8. Liu, Q., Safavi-Naini, R., Sheppard, N.P.: Digital rights management for content distribution. In: Proceedings of the Australasian information Security Workshop Conference on ACSW Frontiers 2003. (2003) 49–58
9. Feigenbaum, J., Freedman, M., Sander, T., Shostack, A.: Privacy engineering for digital rights management systems. In: Proceedings of the 2001 ACM Workshop on Security and Privacy in Digital Rights Management. (2001) 76–105
10. Clip2: The Gnutella Protocol Specification v0.4. (2001)
11. Aberer, K., Cudre-Mauroux, P., Datta, A., Hauswirth, M.: PIX-Grid: A platform for P2P photo exchange. In: Proceedings of Ubiquitous Mobile Information and Collaboration Systems (UMICS 2003), collocated with CAiSE'03. (2003)
12. Turcan, E., Shahmehri, N., Graham, R.L.: Intelligent software delivery using P2P. In: IEEE Proceedings of Second International Conference on Peer-to-Peer Computing. (2002) 136–143
13. Sun Microsystems: Project JXTA v2.0: $Java^{TM}$ Programmer's Guide. (2003)

How Social Structure Improves Distributed Reputation Systems - Three Hypotheses[*]

Philipp Obreiter, Stefan Fähnrich, and Jens Nimis

Institute for Program Structures and Data Organization,
Universität Karlsruhe (TH),
D-76128 Karlsruhe, Germany
{obreiter, nimis}@ipd.uni-karlsruhe.de
stefan@faehnrich.de

Abstract. Reputation systems provide an incentive for cooperation in artificial societies by keeping track of the behavior of autonomous entities. The self-organization of P2P systems demands for the distribution of the reputation system to the autonomous entities themselves. They may cooperate by issuing recommendations of other entities' trustworthiness. The recipient of a recommendation has to assess its truthfulness before taking it into account. The current assessment methods are based on plausibility considerations that have several inherent limitations. Therefore, in this paper, we propose social structure as a means of overcoming some of these limitations. For this purpose, we examine the properties of social structure and discuss how distributed reputation systems can make use of them. This leads us to the formulation of three hypotheses of how social structure overcomes the limitations of plausibility considerations. In addition, it is pointed out how the hypotheses can be tested.

1 Introduction

In general, autonomous agents lack inherent incentives to exhibit cooperative behavior in open artificial societies. The predominance of uncooperative behavior leads to the emergence of a market for lemons [1] that, in turn, deprives cooperative agents of incentives for participation. Therefore, it is crucial to provide incentives for cooperation. The analysis of the design space of such incentives [2] has shown that the enforceability of every incentive pattern eventually relies on a reputation system. A reputation system keeps track which agents keep their promises and which do not. This information is exploited in order to pre-estimate the behavior of potential transaction partners. By this means, a reputation system supports the trust formation process of each agent.

The application of reputation systems to self-organized P2P systems appears especially challenging. In the absence of any commonly trusted entity, the reputation system has to be distributed to the autonomous agents themselves. They

---

[*] The work done for this paper is partly funded by the German Research Community (DFG) in the context of the priority programs (SPP) no. 1140 and no. 1083.

G. Moro, S. Bergamaschi, and K. Aberer (Eds.): AP2PC 2004, LNAI 3601, pp. 229–236, 2005.

have to make up for the lack of centralized reputation processing by exchanging recommendations of other agents' trustworthiness [3] . The major challenge of distributed reputation systems consists in assessing the truthfulness of such recommendations. In this regard, a recommendation is truthful if it corresponds to the experiences made by the recommender.

The existing approaches for distributed reputation systems (e.g., [4,5]) make use of plausibility considerations in order to provide for such assessment. This means that the impact of a recommendation is contingent upon its plausibility which, in turn, depends on its compatibility with prior beliefs. Such schemes are vulnerable to misbehavior that is aimed at influencing the plausibility considerations themselves [6]. In the human society, plausibility considerations are complemented with considerations based on social structuring in order to increase the accuracy of assessing recommendations [7,8,9]. This observation leads us to propose the application of social structure to distributed reputation systems for artificial societies.

The paper is structured as follows: In Section 2, we analyze distributed reputation systems and point out the limitations of existing approaches to assess the truthfulness of recommendations. In Section 3, we examine the properties of social structure and the means of making use of them in distributed reputation systems. We formulate three hypotheses of how social structure improves distributed reputation systems in Section 4. The paper is concluded in Section 5.

2   Distributed Reputation Systems

In our previous work [6], we provided an analysis of distributed reputation systems and proposed the use of non-repudiable tokens (so-called evidences) in order to improve their effectiveness. In this section, we recapitulate the findings of that work and point out the need for a complementary means of increasing the effectiveness of distributed reputation systems.

System Model and Requirements. We assume a system that consists of autonomous *agents* that may exchange messages through authenticated channels. Each agent runs a *local instance* of the reputation system. These instances may cooperate by exchanging *recommendations*. The issuer of a recommendation (*recommender*) communicates the trustworthiness of a certain agent to the *recipient* of the recommendation. Recommendations may be untruthful. Therefore, the recipient has to assess the truthfulness of the recommendation before taking it into account. In the following, an agent that performs such assessment will be referred to as *assessor*.

Limitations of Plausibility Considerations. Plausibility considerations are contingent upon prior beliefs. More specifically, the considerations comprise two parts. On the one hand, a recommendation is assessed as rather trustworthy if it is compatible to the first hand experiences made by the assessor itself. On the other hand, the more the recommender is trusted the more the recommendation is regarded as truthful. In [6], we identified and discussed the limitations of plausibility considerations. In the following, we focus on three limitations that are addressed by the hypotheses of Section 4.

1. **Support for newcomers:** If an agent enters into the system for the first time, he lacks first hand experiences with other agents. In addition, the newcomer does not know which agent can be trusted as a recommender. Consequently, newcomers are not able to evaluate the plausibility of recommendations.
2. **Recognition of praising:** Agents are able to form a collusion by mutually overstating their trustworthiness. The colluding agents may render their mutual recommendations plausible by truthfully recommending entities that are not member of the collusion. Hence, such praising cannot be fully identified by the means of plausibility considerations.
3. **Dissemination of recommendations:** Self-recommendations are needed in situations in which the agents that are aware of the good conduct are offline or unwilling to issue recommendations about the respective agent. If self-recommendations are infeasible in such circumstances, the dissemination of recommendations is not *effective*. From the point of view of *efficiency*, the use of self-recommendations would also be desirable since an agent would be able to self-recommend on demand whenever a transaction is imminent. However, the use of plausibility considerations implies that a self-recommendation can only be credibly communicated if its issuer has a good reputation. Yet, in such a case, the other agents would already be aware of the good conduct of the self-recommender so that the self-recommendation becomes dispensable. Therefore, self-recommendations are not meaningful if plausibility considerations are applied. As a result, the dissemination of recommendations is neither effective nor efficient.

Limitations of Evidence Awareness. In [6], it is proposed to make use of evidences in order to overcome the limitations of plausibility considerations. An *evidence* is a non-repudiable token that may be arbitrarily transferred[1]. In contrast to plausibility considerations, the application of evidences achieves a coupling between the actual behavior and the assessment of recommendations about it. This is because the assessment of such behavior is conducted as a verification process that is solely based on evidences. Yet, the coupling is not perfect since there are inherent restrictions for the issuance of evidences. If behavior is not documented by evidences, it has still to be assessed by the means of plausibility considerations. Therefore, some limitations of plausibility considerations cannot be fully overcome.

For the three limitations that we consider in this paper, the application of evidences translates as follows: *Newcomers* are better supported since they do not have to resort to plausibility considerations if evidence based verification is possible. However, they still lack support for the assessment of undocumented behavior. Furthermore, evidences do not provide a means of coping with *praising*.

---

[1] The term evidence has been used differently in reputation systems. In [3,10], it depicts witnessed circumstances, i.e., first hand experiences and recommendations.

This is because the colluders could agree to mutually issue evidences of their good conduct. Finally, the transferability of evidences allows for *disseminating* self-recommendations. Yet, agents will provide a one-sided view of their track record by only including evidences of his good behavior. However, the number positive evidences does not give information about the ratio of good and bad behavior of the self-recommender. Therefore, it is unclear to which extent such one-sidedness compromises the meaningfulness of self-recommendations.

We conclude that evidence awareness has to be complemented by a further approach in order to overcome the limitations of plausibility limitations.

3   Social Structure

In this section, we present social structure as an approach that copes with the remaining limitations of plausibility limitations. For this purpose, we define the concept of social structure and examine its properties. Subsequently, we point out how social structure can be applied in distributed reputation systems and discuss existing approaches that pursue this idea.

3.1   Definition and Properties

According to Giddens [9], social structure is defined as "the patterning of interaction, as implying relations between actors or groups, and the continuity of interaction in time". From this definition, we deduce the following two main aspects:

– **Relationships:** The patterning of interaction implies that an agent may establish a relationship with some of the other agent. Such a relationship has a certain type and does not have to be mutual in nature. In the following, we refer to a relationship of an agent A with an agent B as $A \longrightarrow_{type} B$. Relationships are either *directed* (e.g., $Child \longrightarrow_{obedience} Parent$) or *mutual* (e.g., $Child \longleftrightarrow_{reliance} Parent$). The introduction of relationships facilitates the formation of groups. In this regard, a *group* is a set of agents that have mutual relationships of the same type.
– **Dynamics:** The temporal notion of the definition implies that relationships are not necessarily *pre-defined*. They might as well be *adaptive* such that the relationship network is responsive to time-variant cooperation patterns. In such a case, there have to be criteria of when a relationship is to be *established* and *cancelled*. If a relationship is directed, each agent is able to test these criteria and perform the establishment and cancellation of relationships on himself. However, for mutual relationships, the agents have to coordinate the establishment and cancellation of their relationship. Therefore, mutual relationships are more difficult to maintain than directed ones.

3.2   Application to Distributed Reputation Systems

We propose to apply social structure to distributed reputation systems in order to overcome the limitations of plausibility considerations. For this purpose, we examine how a distributed reputation system can make use of the properties of social structure.

There are several types of relationships that make sense in the context of a distributed reputation system. The *trust*-relationship and *distrust*-relationship refer to high and low levels of trust respectively. A more specific type of relationship is the *bail*-relationship. If an agent (*bailor*) establishes such a relationship to another agent (*bailee*), he agrees to be punished for the misbehavior of the bailee. Therefore, the establishment of a bail-relationship necessitates especially high trust levels by the bailor.

Based on these types of relationships, the distributed reputation system can be extended as follows:

- **Dissemination of recommendations:** If an agent wants to evaluate trustworthiness of other agents, he prefers to ask those agents with whom he has a trust-relationship. For the pro-active dissemination of one's own recommendations, the recommender has to know which agents have a trust-relationship with himself, as it is the case for mutual trust-relationships.
- **Assessment of recommendations:** Upon receival of a recommendation, an agent has to assess its truthfulness. Such assessment can be based on his relationship with the recommender. This means that the assessor perceives the recommendation as more truthful if he has a trust-relationship with the recommender.
- **Self-recommendations:** An agent is able to self-recommend by stating which agents have a bail-relationship with him. The more bailors an agent has, the more he appears trustworthy. The assessor of such a self-recommendation has to be convinced that the stated agents are indeed bailors. There are two means of doing so. First, the assessor may request that the stated agents confirm their status as bailors. Yet, the agents might be offline or not willing to respond to such request. Second, a bailor may render the bail-relationship non-repudiable by handing over an evidence of the relationship to the bailee. In such a case, a self-recommender is able to prove that the stated agents are indeed ready to bail for him. The disadvantage of this approach is that bail-relationships either cannot be cancelled or are temporally bounded. In the latter case, bail-evidences have to be issued repeatedly.

3.3   Existing Approaches

Among the existing distributed reputation systems [4,5,11,12], only few make use of social structure. In the following, we take a closer look at them.

Friends&Foes [11] is a distributed reputation system that makes use of an *adaptive* social structure[2]. It consists of two types of *directed* relationships, i.e., the *trust*-relationship (so-called friends) and the *distrust*-relationship (so-called foes. These relationships make the one-sided trust levels explicit. Yet, they are

---

[2] A further distributed reputation system based on the notion of friends is presented in [5]. It applies directed trust-relationships but fails to make them explicit. Hence, the contents of recommendations cannot be based on social structure.

only exploited for the formation of trust. Hence, the dissemination and assessment of recommendations does not make use of the social structure.

The *Buddy System* [12] is a further distributed reputation system that applies social structure. The only type of relationship is a mutual *buddy*-relationship that is established adaptively between a pair of agents (so-called buddies). A *buddy*-relationship is a combination of the *trust*-relationship and the *bail*-relationship. There are two criteria for its establishment. Apart from mutually trusting each other, the agents have to perceive the trustworthiness of other agents likewise. This additional criterion is called similarity of world views. It is set in place in order to reduce the conflict potential between buddies. Buddy-relationships are not documented with evidences. Therefore, the assessor of a self-recommendation probabilistically contacts some of the alleged bailors. Based on their responses, the assessor makes a projection of the actual number of buddies. Recommendations about third parties can also be issued in the Buddy System. The recommender disseminates them to his buddies.

4   The Impact of Social Structure: Three Hypotheses

In this section, we formulate three hypotheses of how social structure overcomes the limitations of plausibility considerations. Furthermore, we point out the intuition that lies behind each hypothesis and discuss for each hypothesis how it can be tested in the context of simulations.

> **Hypothesis 1: Orientation for Newcomers**
> Social structure provides an orientation for newcomers such that they are able to assess the trustworthiness of other agents.

The intuition behind this hypothesis is that the trustworthiness of agents is reflected in the relationship network of the social structure. Hence, it suffices for a newcomer to gain an overview of the relationship network in order to find out about the trustworthiness of the other agents. For this purpose, a newcomer has to request self-recommendations from those agents he is interested in.

For the test of this hypothesis, the performance of newcomers has to be compared for a distributed reputation system with and without social structure. In this respect, performance refers to the portion of successful transactions, i.e., transactions without defection. The hypothesis can be validated if the performance experiences a sustained amelioration by the presence of social structure.

> **Hypothesis 2: Protection Against Collusions**
> Social structure curbs the impact of colluding agents that mutually praise.

In the presence of social structure, colluders have to choose among two options. If they opt to make their collusion explicit by establishing mutual bail-relationships, the collusion becomes identifiable. For this purpose, it suffices to gain bad experiences with few of the colluders in order to conjecture that the group is indeed a collusion. However, if the colluders refrain from establishing relationships, conventional recommendations provide the only means of mutual praising. In such a case, their recommendations tend to be dwarfed by the bail-relationships that well-behaving agents establish. Therefore, regardless of which option the colluders choose, the impact of praising is marginalized or praising even becomes counter-productive.

The hypothesis can be tested by comparing the performance of colluding and non-colluding agents in the presence of social structure. Such comparison has to distinguish between colluders who opt for making their collusion explicit and those who resort to conventional recommendations.

Hypothesis 3: Effective and Efficient Dissemination
Social structure allows for a more effective and efficient dissemination of recommendations.

The intuition behind this hypothesis is twofold. First, recommendations may be disseminated more convincingly along the relationship network of the social structure. By disseminating along the trust-relationships, a recommendation is only obtained by those agents that take it most into account. Therefore, we are able to economize on the dissemination to agents that would neglect the recommendation. This yields an increased efficiency of disseminating recommendations. Second, an agent may self-recommend by stating which agents are ready to bail for him. According to Section **??**, the possibility to self-recommend implicates more effectiveness and efficiency for the dissemination of recommendations.

For the test of this hypothesis, we have to show that a distributed reputation system with social structure dominates any plausibility based distributed reputation system in terms of both effectiveness and efficiency. The *effectiveness* of disseminating recommendations can be evaluated by measuring the effectiveness of the distributed reputation system itself. In [13], the coefficient of correlation between individual cooperation costs and utilities has been proposed as such a measure of effectiveness. More specifically, the higher the coefficient of correlation and the higher the slope of the regression line, the more effective the distributed reputation system is. The *efficiency* of disseminating recommendations can be evaluated in a more straightforward manner by keeping track of the number of sent recommendation messages.

5   Conclusion

Distributed reputation systems provide an incentive for cooperation in P2P systems by keeping track of the behavior of autonomous entities. The agents may cooperate by issuing recommendations of other entities' trustworthiness. Currently, the assessment of such recommendations is based on plausibility considerations and evidence awareness. In this paper, we have pointed out that these assessment methods require a further means of dealing with their shortcomings. For this purpose, we have proposed that social structure is applied to the distributed reputation system. We have examined the properties of social structure and discussed its impact on distributed reputation systems. This led us to the formulation of three hypotheses of how social structure overcomes the limitations of plausibility considerations. We have pointed out how the hypotheses can be tested.

In the future, we will test the hypotheses by the means of simulation. The tentative simulation results[3] corroborate the first two hypotheses and indicate the validity of the third one. Furthermore, we plan to investigate whether there is a generic means of augmenting existing distributed reputation systems with social structure.

References

1. Akerlof, G.: The market for lemons: Quality uncertainty and the market mechanism. Quarterly Journal of Economics **89** (1970) 488–500
2. Obreiter, P., Nimis, J.: A taxonomy of incentive patterns - the design space of incentives for cooperation. In: Second Intl. Workshop on Agents and Peer-to-Peer Computing (AP2PC'03), Springer LNCS 2872, Melbourne, Australia (2003)
3. English, C., Wagealla, W., Nixon, P., Terzis, S., Lowe, H., McGettrick, A.: Trusting collaboration in global computing systems. In: Proc. of the First Intl. Conf. on Trust Management (iTrust), Heraklion, Crete, Greece (2003) 136–149
4. Kinateder, M., Rothermel, K.: Architecture and algorithms for a distributed reputation system. In Nixon, P., Terzis, S., eds.: Proc. Of the First Intl. Conf. On Trust Management (iTrust), Heraklion, Greece, Springer LNCS 2692 (2003) 1–16
5. Marti, S., Garcia-Molina, H.: Limited reputation sharing in P2P systems. In: ACM Conference on Electronic Commerce (EC'04). (2004)
6. Obreiter, P.: A case for evidence-aware distributed reputation systems. In: Second International Conference on Trust Management (iTrust'04), Oxford, UK, Springer LNCS 2995 (2004) 33–47
7. Myers, D.G.: Psychology. 6th edn. Worth Publishers (2001)
8. Hogg, M.A.: Intragroup processes, group structure and social identity. In Robinson, W., ed.: Social Groups and Identities: Developing the Legacy of Henri Taiffel, Butterworth Heinemann (1996)
9. Giddens, A.: The Constitution of Society: Outline of a Theory of Social Structuration. Polity Press, Cambridge, MA (1984)
10. Yu, B., Singh, M.P.: An evidential model of distributed reputation management. In: Proceedings of the First International Joint Conference on Autonomous Agents and Multiagent Systems (AAMAS'02), Bologna, Italy (2002) 294–301
11. Miranda, H., Rodrigues, L.: Friends and foes: Preventing selfishness in open mobile ad hoc networks. In: Proc. of the First Intl. Workshop on Mobile Distributed Computing (MDC'03), Providence, RI, USA, IEEE Computer Society Press (2003)
12. Fähnrich, S., Obreiter, P., König-Ries, B.: The buddy system: A distributed reputation system based on social structure. In: 7th Intl. Workshop on Data Management in Mobile Environments, Ulm (2004)
13. Obreiter, P., König-Ries, B., Papadopoulos, G.: Engineering incentive schemes for ad hoc networks - a case study for the lanes overlay. In: First EDBT-Workshop on Pervasive Information Management, LNCS 3268. (2004)

---

[3] We discuss the results at http://www.ipd.uka.de/~obreiter/publications.html

Opinion Filtered Recommendation Trust Model in Peer-to-Peer Networks*

Weihua Song and Vir V. Phoha

Computer Science, College of Engineering and Science,
Louisiana Tech University, Ruston, LA 71272, USA
{wso003, phoha}@latech.edu

Abstract. A multiagent distributed system consists of a network of heterogeneous peers of different trust evaluation standards. A major concern is how to form a requester's own trust opinion of an unknown party from multiple recommendations, and how to detect deceptions since recommenders may exaggerate their ratings. This paper presents a novel application of neural networks in deriving personalized trust opinion from heterogeneous recommendations. The experimental results showed that a three-layered neural network converges at an average of 12528 iterations and 93.75% of the estimation errors are less than 5%. More important, the model is adaptive to trust behavior changes and has robust performance when there is high estimation accuracy requirement or when there are deceptive recommendations.

1 Introduction

In a system where no global or central reputation mechanism is available, an alternative is to aggregate a collection of local trust evaluations, i.e., recommendations. Current recommendation trust models differ in their selections of recommenders and in their aggregations of recommendations. Examples are social network topological model by Pujol et al. [1], Bayesian rating model by Mui et al. [2], Bayesian Network by Y. Wang et al. [3], and Dempster-Shafer belief model by Y. Bin et al. [4] etc.

Zacharia and Mae's HISTO model [5] applies raters' reputations and the deviation of the ratings as weights to the recommended trust value. It uses breadth first search algorithm to find all the referral chains within certain length limit and branching size. Riggs et al. [6] developed a quality filtering trust model. The model rates reviewers and applies the quality of the reviewers into merits of the reviewed papers. Mui et al. [2] proposed a recommendation trust model based on Beyesian probability theory. A referral chain's trust is measured by successive applications of Beyesian probability of two contiguous references along the chain. Chernoff Bound is used as a reliability measurement for trust information gathered along each chain. Yu et al. [4] presented an evidence model

---

* This work is supported in part by the Army Research Office under Grant No. DAAD 19-01-1-0646 and by Louisiana Board of Regents under Grant LEQSF(2003-05)-RD-A-17.

G. Moro, S. Bergamaschi, and K. Aberer (Eds.): AP2PC 2004, LNAI 3601, pp. 237–244, 2005.

to evaluate recommendation trust. The model applies Dempster-Shafer theory to multiple witnesses. Pujol et al. [1] proposed a recommendation trust model that applies noderanking algorithm to infer a node's reputation based on social network topology.

However, there are a few caveats to the approaches mentioned above. First, different users may arrive at significantly varying estimates of performance of the same service provider. Second, different users may be able to observe different instances of the performance of a given service provider. Third, deceptive recommendations may exist. More sophisticated collaborative trust models [3][7][8][9][10][11][12] are developed.

Wang et al. [3] applied Naive Bayesian network to recommendation trust since trust is multifaceted. EigenTrust model of Kamvar et al. [10] focuses on detecting malicious file providers in peer-to-peer file sharing networks. The model is built on the notion of transitive trust, i.e., a peer trusts authentic file providers as well as their recommendations. Yu et al. [12] applied weighted majority technique to belief function and belief propagation. The model assigns weights to recommenders and decreases the weights assigned to unsuccessful recommenders.

Our approach is a novel application of neural network techniques in recommendation trust. It derives a requester's own trust opinion from multiple heterogeneous recommendations, i.e., recommendations of different trust estimation processes and evaluation models with or without deceptions. It concentrates on the algorithms of selecting *qualified recommenders* and propagating their recommendations into the requester's own trust opinion. The approach assumes that a requester's trust opinion v_0 is a function of M recommenders' recommendations v_i, $i = 1, 2, \ldots, M$:

$$v_0 = F(v_1, v_2, \ldots, v_M) \ . \tag{1}$$

What we do is to find out those M recommenders, i.e., *qualified recommenders*, and to approximate function F. Suppose there are N movie file providers with whom the requester and the M recommenders all have direct experiences. Given the requester's and the recommenders' trust opinions of N providers, v_i^j, where $i = 0, 1, \ldots, M, j = 1, 2, \ldots, N$, our work is to find a function F such that the summation of the squared estimation errors is minimized.

$$min \sum_{j=1}^{N} (v_0^j - F(v_1^j, v_2^j, \ldots, v_M^j))^2 \ . \tag{2}$$

Where v_0^j stands for the requester's trust opinion of movie file provider j, and v_i^j stands for recommender i's trust opinion of movie file provider j. Once F is found, we can plug in the M recommenders' trust opinions of the party of interest and obtain immediately the requester's trust estimation of the party of interest.

The rest of the paper is organized as follows. Section 2 develops an ordered depth-first recommendation trust network and designs an algorithm to identify *qualified recommenders*. Section 3 introduces a neural network-based recommendation trust model. Section 4 presents experimental results. Section 5 concludes the paper.

2 Ordered Depth-First Recommendation Network and Qualified Recommenders

An ordered depth-first search for current recommenders is developed. Among current recommenders, we select *qualified recommenders*, whose trust opinions are used to build the opinion filtered neural network trust model (OFNN).

2.1 Ordered Depth-First Recommendation Network

A requester keeps a rated recommender set per trust context. A recommender set R contains a *qualified recommender* set QR and an unqualified recommender set NQR. All qualified recommenders are ranked higher than unqualified recommenders. *Qualified recommenders* are ranked by the number of times they have been selected as qualified recommenders. Unqualified recommenders are ranked by the number of times they have been selected as recommenders but not qualified. Initially, the *qualified recommender* set is empty and the unqualified recommender set consists of all the acquaintances of the requester. The acquaintances are ranked by their trust values. R is updated and reordered after the requester sends a new query to every $r \in R$ and obtains a current set of *qualified recommenders* QR_c and unqualified recommenders NQR_c.

We use ordered depth-first search (ODFS) for current recommenders R_c. R_c is made up of those having direct trust experience with the party of interest. The depth-first search is ordered by recommenders' rates. That is, a highly rated recommender's trust opinion is processed first. We set limits on chain length and branching size of an ODFS. An ordered depth-first recommendation network need building only when the qualified recommenders of an OFNN are unable to provide recommendations for the current trust query.

2.2 Identification of Qualified Recommenders

Qualified recommenders are those in current recommender set R_c and having direct experiences with a set of parties (under the same trust context) that the requester also has direct experiences with. We assume peers are willing to exchange trust opinions for mutual benefits. To identify qualified recommenders QR_c and those parties(movie file providers in our case), a requester first selects top N active movie providers P with whom he has direct experiences. The requester exchanges trust opinions of providers P with current recommenders R_c. A two dimensional array RP is built where element $RP[i][j]$ is either 1 or 0, representing that recommender $r_i \in R_c$ has or has not direct trust experiences with movie file provider $p_j \in P$. Recommenders that know less than $ceil_1$ movie file providers are excluded and movie providers that are known by less than $ceil_2$ recommenders are also excluded. A new RP array is built based on the selected recommenders and movie file providers. The top $ceil_3$ recommenders that know the majority of the movie file providers are selected as qualified recommenders. Those movie file providers that are known by all the $ceil_3$ qualified recommenders forms new P.

3   Opinion Filtered Recommendation Trust Model

There is an OFNN per trust context per peer. A peer(requester)'s OFNN under certain trust context is trained by his and his qualified recommenders' trust opinions, for example the trust opinions of movie file providers P. The input layer of the OFNN contains as many neurons as the number of qualified recommenders. The input neurons represent the qualified recommenders' trust opinions. The output layer contains only one neuron, representing the requester's trust opinion. There can be as many hidden layers and hidden neurons as necessary to catch the nonlinear relationship between input and output values and to satisfy the model's convergent speed. We set up a three-layered OFNN with random initial neuron connections and a random learning rate. We then apply the backpropagation algorithm [13] to train the OFNN. The details are as the following. Qualified recommenders' trust opinions flow forward through the neural network as:

$$o_u = \frac{1}{1 + e^{-\boldsymbol{W}_u \boldsymbol{R}_u}} \tag{3}$$

Approximation errors flow backward at the output neuron and at the hidden neurons as:

$$\delta = o(1 - o)(t - o) \tag{4}$$

$$\delta_h = o_h(1 - o_h)w_h\delta \tag{5}$$

Neuron connections are updated as:

$$w_{ij} = w_{ij} + \eta\delta_i o_i \tag{6}$$

Where $\boldsymbol{W}_u$ and $\boldsymbol{R}_u$ represent a weight vector and a reputation vector forwarded to neuron u respectively. o is the estimated reputation opinion generated by the output neuron, and o_i is the output value of neuron i. t is the requester's reputation opinion. w_h is the weight of a connection between hidden neuron h and the output neuron. w_{ij} is the weight from neuron i to neuron j. δ and δ_h stand for error items at the output neuron and hidden neuron h respectively. Algorithm 1 describes the entire process. The algorithm can be optimized by storing several most recently used neural networks instead of one last updated neural network. There is a trade off between memory (storage of more than one neural networks per trust context) and computation speed (time spent on finding R_c, identifying QR_c and retraining the OFNN).

Algorithm 1. Build Opinion Filtered Trust Model

1: if (qualified recommenders of the OFNN can not provide recommendations)
2:     ordered depth-first search for recommenders R_c;
3:     identify qualified recommenders QR_c;
4:     train Neural Network by backpropagation algorithm;
5: end if
6: update rated recommender set R;
7: input recommendations of the current request through the neural network;
8: output the personalized opinion filtered recommendation trust

4   Experiment Results

We simulated trust behavior of 50 agents in a P2P movie file sharing network. Total 500 transactions were simulated. Movie file providers and movie file downloaders were generated randomly from the 50 agents. A movie file provider's trust behavior was evaluated by two factors, file quality and download speed of the file. Each agent had an average download speed quality $\overline{u}_1$ and an average file quality $\overline{u}_2$. $\overline{u}_1$ and $\overline{u}_2$ were randomly generated in a range of $[0, 1]$. The actual download quality u_1 and download speed u_2 of a transaction were generated by identical and independent distributions distributed uniformly and randomly centering with mean value $\overline{u}_1$ and $\overline{u}_2$ respectively. A single transaction was evaluated by the weighted average of u_1 and u_2. The normalized weights of those two factors varied from one agent to another. The weights were generated randomly in a range of $[0, 1]$. We also simulated non deceptive and deceptive recommendations. We assume a recommendation follows:

$$v_{rec} = min(1, cv_{act}) \ . \tag{7}$$

Where c is a factor larger than zero, v_{act} is the actual trust rating, and v_{rec} is the recommendation. If c is larger or less than 1, the recommendation is deceptive.

Our experiments were based on the simulated data. We set 10 different estimation error thresholds and ran the opinion filtered recommendation trust model twenty times for each size. To train a trust neural network, we randomly set learning speed η in a range of $[0.4, 0.6]$, and set the initial weights of neuron connections in a range of $[-0.05, 0.05]$. A *correct* estimation is the one satisfying $\frac{|o_i - t_i|}{t_i} \leq \theta$ (θ is a constant), and correctness function T is:

$$T(o_i, t_i) = \begin{cases} 1, \text{ for } \frac{|o_i - t_i|}{t_i} \leq \theta \\ 0, \text{ otherwise} \end{cases} \tag{8}$$

Where θ is an estimation error threshold, o_i and t_i stand for the output trust and the requester's actual trust evaluation of movie file provider p_i individually.

Table 1 shows the average convergent speed and estimation correctness under various estimation error thresholds. The estimations were 100% correct and converged at 404,047 iterations if the estimation errors were allowed to be no

more than 15%. We had 93.8% correct estimations if the estimation accuracy
was 95% and above. Figure 1 shows the performance of the opinion filtered trust
model under different accuracy requirements with and without deceptive recommendations. The model converged at an average of 34.8% more iterations under
deceptive recommendations than under honest recommendations when estimation errors were set less than 15%. When estimation errors were in the range from
15% to 30%, surprisely, the model converged faster by 12.5% under deceptive
recommendations than under honest recommendations. This might result from
differences in randomly generated learning rates and initial weights of neuron
connections. Estimation error threshold $\theta = 10\%$ is a critical point. The convergent speed dramatically slows down if θ is less 10%. Figure 2 compares the
convergent speed and estimation accuracy under 36 and 16 sets of recommendations with deceptions. It demonstrates that the model provided comparable
correctness estimations with 6.4% less iterations under 36 sets of recommendations .

Table 1. Convergence and Correctness of Opinion Filtered Trust Model

Estimation Error θ	Learning Rate η	Convergence (number of iterations)	Correctness $\sum_i^{16} T(o_i, t_i)$
30%	0.485	54042	16
25%	0.48	165260	16
20%	0.49	273207	16
15%	0.515	404047	16
10%	0.555	551989	15
9%	0.475	684838	15
8%	0.5	810923	15
7%	0.59	1006285	15
6%	0.595	1221718	15
5%	0.595	1435770	15

5   Conclusion

This paper studies the problem of heterogeneous and deceptive recommendations
in trust management. It focuses on how to accurately and effectively derive trust
value of an unknown party from multiple recommendations. It designs an ordered
process of depth-first search for recommenders. It also develops an algorithm to
identify qualified recommenders and to aggregate their recommendations. The
neural network trust model helps an agent to minimize the effect of deceptions
and heterogeneous trust evaluation standards. The model derives a trust value
based on an agent's own trust standards and thus makes trust decision easier.
The experimental results show that the neural network model converges at fast
speed with high accuracy. More important, the model has good performance
under various accuracy requirements and is capable to aggregate multiple recommendations non-linearly. For simplicity, the experiments were carried out based

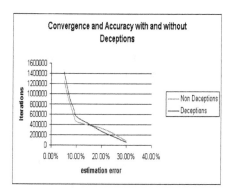
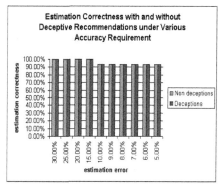

(a) Comparison of convergent speed

(b) Comparison of estimation correctness

Fig. 1. Performance of the trust model with and without deceptions under various estimation accuracy requirement

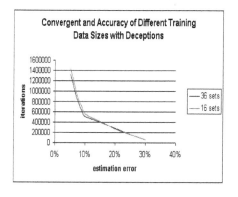
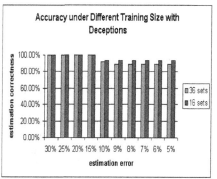

(a) Comparison of convergent speed

(b) Comparison of estimation correctness

Fig. 2. Performance of the trust model trained by 16 and 36 sets of recommendations with deceptions under various estimation accuracy requirement

on a simple deception model and on weighted average trust estimation processes. Conceptually, the model can be applied to more complicated deceptions and to different estimation processes and local trust models. The model is adaptive to changes of agents' trust behaviors and expertise as well. This can be easily achieved by retraining a recommendation trust neural network.

In future work, we plan to integrate the model in a dynamic environment of a multiagent system, where agents adapt their strategies and trust models. We also plan to study a recommender's reputation and compare various trust models' performance.

References

1. Pujol, J., Sang, R., Halberstadt, A.: Extracting reputation in multi agent systems by means of social network topology. In: Proceedings of the first International Joint Conference on Autonomous Agents and Multiagent Systems (AAMAS02). (2002)
2. Mui, L., Mohtashemi, M., Halberstadt, A.: A computational model for trust and reputation. In: Proc. 35th Hawaii International Conference on System Sciences. (2002)
3. Wang, Y., Vassileva, J.: Beyesian network-based trust model. In: Proc. of IEEE/WIC International Conference on Web Intelligence (WI2003). (2003)
4. Yu, B., Singh, M.P.: Towards a probabilistic model of distributed reputation management. In: Proc. 4th Workshop on Deception, Fraud and Trust in Agent Societies. (2001)
5. Zacharia, G., Mae, P.: Collaborative reputation mechanisms in electronic marketplaces. In: Proc. 32nd Hawaii International Conf on System Sciences. (1999)
6. Riggs, T., Wilensky, R.: An algorithm for automated rating of reviewers. In: Proceedings of first ACM and IEEE Joint Conference on Digital Libraries (JCDL01). (2001)
7. Azzedin, F., Maheswaran, M.: Evolving and managing trust in grid computing systems. In: Proceedings of IEEE Canadian Conference on Electrical & Computer Engineering (CCECE02). (2002)
8. Daniani, E., Vimercati, S., Paraboschi, S., Samarati, P., Violante, F.: A reputation-based approach for choosing reliable resources in peer-to-peer networks. In: Proceedings of the 9th ACM Conference on Computer and Communications Security (CCS02). (2002)
9. Gupta, M., Judge, P., Ammar, M.: A reputation system for peer-to-peer networks. In: Proceedings of the 13th ACM International workshop on Network and Operating Systems Support for Design Audio and Vedio (NOSSDAV03). (2003)
10. Kamvar, S., Schlosser, M., Garcia-Molina, H.: The eigentrust algorithm for reputation management in p2p networks. In: Proceedings of the twelfth ACM International World Wide Web Conference (WWW03). (2003)
11. Schafer, J.B., Konstan, J., J.Riedl: Meta-recommendation systems: User-controlled integration of diverse recommendations. In: Proceedings of ACM Conference on Information and Knowledge Management (CIKM02). (2002)
12. Yu, B., Singh, M.P.: Detecting deception in reputation management. In: AAMAS'03. (2003)
13. T.Mitchell: Machine Learning. McGraw-Hill (1997)

Author Index

Ai, Changquan, 193
Amamiya, Makoto, 62

Bergamaschi, Sonia, 120
Boella, Guido, 86
Börner, Katy, 14

Chen, Jih-Yin, 221
Chiang, Chih-He, 221
Chun, Brent, 28
Cortés, Ulises, 98
Cranefield, Stephen, 153
Crespo, Arturo, 1
Cruz, Isabel F., 108

Dasgupta, Prithviraj(Raj), 213
Dogdu, Erdogan, 144
Dury, Arnaud, 185

Fähnrich, Stefan, 229
Fillottrani, Pablo R., 120
Fletcher, George H.L., 14

Garcia-Molina, Hector, 1
Garside, Noel, 153
Gelati, Gionata, 120

Hsieh, Chun-Wei, 221
Hsu, Feihong, 108
Hsu, Jane Yung-jen, 221
Huang, Ting-Shuang, 221
Hu, Zhengguo, 193

Jin, Xiaolong, 173

Kogo, Akihiro, 62
Koyanagi, Keiichi, 161
Krishnamoorthy, Savitha, 40

Lauria, Mario, 40
Liu, Jiming, 173

Madiraju, Praveen, 144
Marshall, Lindsay, 54

Matsuno, Daisuke, 62
Mine, Tsunenori, 62

Nimis, Jens, 229
Nowostawski, Mariusz, 153

Obreiter, Philipp, 229
Oliveira, Marcos De, 153

Periorellis, Panayiotis, 54
Phoha, Vir V., 237
Pitsilis, Georgios, 54
Pitt, Jeremy, 74
Prasad, Sushil K., 144
Pujol, Josep M., 98
Purvis, Martin, 153

Ramirez-Cano, Daniel, 74
Regli, William, 201
Regli, William C., 132

Sheth, Hardik A., 14
Song, Weihua, 237
Sultanik, Evan A., 132
Sunderraman, Rajshekhar, 144

Tang, Yan, 193
Thomas, Michael, 201
Torre, Leendert van der, 86
Tsuchiya, Takeshi, 161

Vahdat, Amin, 28
Vaidyanathan, Karthikeyan, 40

Willmott, Steven, 98

Xiao, Huiyong, 108

Yang, Zhen, 173
Yoshikawa, Chad, 28
Yoshinaga, Hirokazu, 161

Zhang, Lin, 193
Zhang, Yang, 193

Lecture Notes in Artificial Intelligence (LNAI)

Vol. 3735: A. Hoffmann, H. Motoda, T. Scheffer (Eds.), Discovery Science. XVI, 400 pages. 2005.

Vol. 3734: S. Jain, H.U. Simon, E. Tomita (Eds.), Algorithmic Learning Theory. XII, 490 pages. 2005.

Vol. 3721: A. Jorge, L. Torgo, P. Brazdil, R. Camacho, J. Gama (Eds.), Knowledge Discovery in Databases: PKDD 2005. XXIII, 719 pages. 2005.

Vol. 3720: J. Gama, R. Camacho, P. Brazdil, A. Jorge, L. Torgo (Eds.), Machine Learning: ECML 2005. XXIII, 769 pages. 2005.

Vol. 3717: B. Gramlich (Ed.), Frontiers of Combining Systems. X, 321 pages. 2005.

Vol. 3702: B. Beckert (Ed.), Automated Reasoning with Analytic Tableaux and Related Methods. XIII, 343 pages. 2005.

Vol. 3698: U. Furbach (Ed.), KI 2005: Advances in Artificial Intelligence. XIII, 409 pages. 2005.

Vol. 3690: M. Pěchouček, P. Petta, L.Z. Varga (Eds.), Multi-Agent Systems and Applications IV. XVII, 667 pages. 2005.

Vol. 3684: R. Khosla, R.J. Howlett, L.C. Jain (Eds.), Knowledge-Based Intelligent Information and Engineering Systems, Part IV. LXXIX, 933 pages. 2005.

Vol. 3683: R. Khosla, R.J. Howlett, L.C. Jain (Eds.), Knowledge-Based Intelligent Information and Engineering Systems, Part III. LXXX, 1397 pages. 2005.

Vol. 3682: R. Khosla, R.J. Howlett, L.C. Jain (Eds.), Knowledge-Based Intelligent Information and Engineering Systems, Part II. LXXIX, 1371 pages. 2005.

Vol. 3681: R. Khosla, R.J. Howlett, L.C. Jain (Eds.), Knowledge-Based Intelligent Information and Engineering Systems, Part I. LXXX, 1319 pages. 2005.

Vol. 3673: S. Bandini, S. Manzoni (Eds.), AI*IA 2005: Advances in Artificial Intelligence. XIV, 614 pages. 2005.

Vol. 3662: C. Baral, G. Greco, N. Leone, G. Terracina (Eds.), Logic Programming and Nonmonotonic Reasoning. XIII, 454 pages. 2005.

Vol. 3661: T. Panayiotopoulos, J. Gratch, R.S. Aylett, D. Ballin, P. Olivier, T. Rist (Eds.), Intelligent Virtual Agents. XIII, 506 pages. 2005.

Vol. 3658: V. Matoušek, P. Mautner, T. Pavelka (Eds.), Text, Speech and Dialogue. XV, 460 pages. 2005.

Vol. 3651: R. Dale, K.-F. Wong, J. Su, O.Y. Kwong (Eds.), Natural Language Processing – IJCNLP 2005. XXI, 1031 pages. 2005.

Vol. 3642: D. Ślezak, J. Yao, J.F. Peters, W. Ziarko, X. Hu (Eds.), Rough Sets, Fuzzy Sets, Data Mining, and Granular Computing, Part II. XXIII, 738 pages. 2005.

Vol. 3641: D. Ślezak, G. Wang, M. Szczuka, I. Düntsch, Y. Yao (Eds.), Rough Sets, Fuzzy Sets, Data Mining, and Granular Computing, Part I. XXIV, 742 pages. 2005.

Vol. 3635: J. Winkler, M. Niranjan, N. Lawrence (Eds.), Deterministic and Statistical Methods in Machine Learning. VIII, 341 pages. 2005.

Vol. 3632: R. Nieuwenhuis (Ed.), Automated Deduction – CADE-20. XIII, 459 pages. 2005.

Vol. 3630: M.S. Capcarrere, A.A. Freitas, P.J. Bentley, C.G. Johnson, J. Timmis (Eds.), Advances in Artificial Life. XIX, 949 pages. 2005.

Vol. 3626: B. Ganter, G. Stumme, R. Wille (Eds.), Formal Concept Analysis. X, 349 pages. 2005.

Vol. 3625: S. Kramer, B. Pfahringer (Eds.), Inductive Logic Programming. XIII, 427 pages. 2005.

Vol. 3620: H. Muñoz-Avila, F. Ricci (Eds.), Case-Based Reasoning Research and Development. XV, 654 pages. 2005.

Vol. 3614: L. Wang, Y. Jin (Eds.), Fuzzy Systems and Knowledge Discovery, Part II. XLI, 1314 pages. 2005.

Vol. 3613: L. Wang, Y. Jin (Eds.), Fuzzy Systems and Knowledge Discovery, Part I. XLI, 1334 pages. 2005.

Vol. 3607: J.-D. Zucker, L. Saitta (Eds.), Abstraction, Reformulation and Approximation. XII, 376 pages. 2005.

Vol. 3601: G. Moro, S. Bergamaschi, K. Aberer (Eds.), Agents and Peer-to-Peer Computing. XII, 245 pages. 2005.

Vol. 3596: F. Dau, M.-L. Mugnier, G. Stumme (Eds.), Conceptual Structures: Common Semantics for Sharing Knowledge. XI, 467 pages. 2005.

Vol. 3593: V. Mařík, R. W. Brennan, M. Pěchouček (Eds.), Holonic and Multi-Agent Systems for Manufacturing. XI, 269 pages. 2005.

Vol. 3587: P. Perner, A. Imiya (Eds.), Machine Learning and Data Mining in Pattern Recognition. XVII, 695 pages. 2005.

Vol. 3584: X. Li, S. Wang, Z.Y. Dong (Eds.), Advanced Data Mining and Applications. XIX, 835 pages. 2005.

Vol. 3581: S. Miksch, J. Hunter, E. Keravnou (Eds.), Artificial Intelligence in Medicine. XVII, 547 pages. 2005.

Vol. 3577: R. Falcone, S. Barber, J. Sabater-Mir, M.P. Singh (Eds.), Trusting Agents for Trusting Electronic Societies. VIII, 235 pages. 2005.

Vol. 3575: S. Wermter, G. Palm, M. Elshaw (Eds.), Biomimetic Neural Learning for Intelligent Robots. IX, 383 pages. 2005.

Vol. 3571: L. Godo (Ed.), Symbolic and Quantitative Approaches to Reasoning with Uncertainty. XVI, 1028 pages. 2005.

Vol. 3559: P. Auer, R. Meir (Eds.), Learning Theory. XI, 692 pages. 2005.

Vol. 3558: V. Torra, Y. Narukawa, S. Miyamoto (Eds.), Modeling Decisions for Artificial Intelligence. XII, 470 pages. 2005.

Vol. 3554: A. Dey, B. Kokinov, D. Leake, R. Turner (Eds.), Modeling and Using Context. XIV, 572 pages. 2005.

Vol. 3550: T. Eymann, F. Klügl, W. Lamersdorf, M. Klusch, M.N. Huhns (Eds.), Multiagent System Technologies. XI, 246 pages. 2005.

Vol. 3539: K. Morik, J.-F. Boulicaut, A. Siebes (Eds.), Local Pattern Detection. XI, 233 pages. 2005.

Vol. 3538: L. Ardissono, P. Brna, A. Mitrovic (Eds.), User Modeling 2005. XVI, 533 pages. 2005.

Vol. 3533: M. Ali, F. Esposito (Eds.), Innovations in Applied Artificial Intelligence. XX, 858 pages. 2005.

Vol. 3528: P.S. Szczepaniak, J. Kacprzyk, A. Niewiadomski (Eds.), Advances in Web Intelligence. XVII, 513 pages. 2005.

Vol. 3518: T.B. Ho, D. Cheung, H. Liu (Eds.), Advances in Knowledge Discovery and Data Mining. XXI, 864 pages. 2005.

Vol. 3508: P. Bresciani, P. Giorgini, B. Henderson-Sellers, G. Low, M. Winikoff (Eds.), Agent-Oriented Information Systems II. X, 227 pages. 2005.

Vol. 3505: V. Gorodetsky, J. Liu, V. Skormin (Eds.), Autonomous Intelligent Systems: Agents and Data Mining. XIII, 303 pages. 2005.

Vol. 3501: B. Kégl, G. Lapalme (Eds.), Advances in Artificial Intelligence. XV, 458 pages. 2005.

Vol. 3492: P. Blache, E. Stabler, J. Busquets, R. Moot (Eds.), Logical Aspects of Computational Linguistics. X, 363 pages. 2005.

Vol. 3490: L. Bolc, Z. Michalewicz, T. Nishida (Eds.), Intelligent Media Technology for Communicative Intelligence. X, 259 pages. 2005.

Vol. 3488: M.-S. Hacid, N.V. Murray, Z.W. Raś, S. Tsumoto (Eds.), Foundations of Intelligent Systems. XIII, 700 pages. 2005.

Vol. 3487: J. Leite, P. Torroni (Eds.), Computational Logic in Multi-Agent Systems. XII, 281 pages. 2005.

Vol. 3476: J. Leite, A. Omicini, P. Torroni, P. Yolum (Eds.), Declarative Agent Languages and Technologies II. XII, 289 pages. 2005.

Vol. 3464: S.A. Brueckner, G.D.M. Serugendo, A. Karageorgos, R. Nagpal (Eds.), Engineering Self-Organising Systems. XIII, 299 pages. 2005.

Vol. 3452: F. Baader, A. Voronkov (Eds.), Logic for Programming, Artificial Intelligence, and Reasoning. XI, 562 pages. 2005.

Vol. 3451: M.-P. Gleizes, A. Omicini, F. Zambonelli (Eds.), Engineering Societies in the Agents World V. XIII, 349 pages. 2005.

Vol. 3446: T. Ishida, L. Gasser, H. Nakashima (Eds.), Massively Multi-Agent Systems I. XI, 349 pages. 2005.

Vol. 3445: G. Chollet, A. Esposito, M. Faundez-Zanuy, M. Marinaro (Eds.), Nonlinear Speech Modeling and Applications. XIII, 433 pages. 2005.

Vol. 3438: H. Christiansen, P.R. Skadhauge, J. Villadsen (Eds.), Constraint Solving and Language Processing. VIII, 205 pages. 2005.

Vol. 3430: S. Tsumoto, T. Yamaguchi, M. Numao, H. Motoda (Eds.), Active Mining. XII, 349 pages. 2005.

Vol. 3419: B. Faltings, A. Petcu, F. Fages, F. Rossi (Eds.), Recent Advances in Constraints. X, 217 pages. 2005.

Vol. 3416: M. Böhlen, J. Gamper, W. Polasek, M.A. Wimmer (Eds.), E-Government: Towards Electronic Democracy. XIII, 311 pages. 2005.

Vol. 3415: P. Davidsson, B. Logan, K. Takadama (Eds.), Multi-Agent and Multi-Agent-Based Simulation. X, 265 pages. 2005.

Vol. 3403: B. Ganter, R. Godin (Eds.), Formal Concept Analysis. XI, 419 pages. 2005.

Vol. 3398: D.-K. Baik (Ed.), Systems Modeling and Simulation: Theory and Applications. XIV, 733 pages. 2005.

Vol. 3397: T.G. Kim (Ed.), Artificial Intelligence and Simulation. XV, 711 pages. 2005.

Vol. 3396: R.M. van Eijk, M.-P. Huget, F. Dignum (Eds.), Agent Communication. X, 261 pages. 2005.

Vol. 3394: D. Kudenko, D. Kazakov, E. Alonso (Eds.), Adaptive Agents and Multi-Agent Systems II. VIII, 313 pages. 2005.

Vol. 3392: D. Seipel, M. Hanus, U. Geske, O. Bartenstein (Eds.), Applications of Declarative Programming and Knowledge Management. X, 309 pages. 2005.

Vol. 3374: D. Weyns, H. V.D. Parunak, F. Michel (Eds.), Environments for Multi-Agent Systems. X, 279 pages. 2005.

Vol. 3371: M.W. Barley, N. Kasabov (Eds.), Intelligent Agents and Multi-Agent Systems. X, 329 pages. 2005.

Vol. 3369: R.V. Benjamins, P. Casanovas, J. Breuker, A. Gangemi (Eds.), Law and the Semantic Web. XII, 249 pages. 2005.

Vol. 3366: I. Rahwan, P. Moraitis, C. Reed (Eds.), Argumentation in Multi-Agent Systems. XII, 263 pages. 2005.

Vol. 3359: G. Grieser, Y. Tanaka (Eds.), Intuitive Human Interfaces for Organizing and Accessing Intellectual Assets. XIV, 257 pages. 2005.

Vol. 3346: R.H. Bordini, M. Dastani, J. Dix, A.E.F. Seghrouchni (Eds.), Programming Multi-Agent Systems. XIV, 249 pages. 2005.

Vol. 3345: Y. Cai (Ed.), Ambient Intelligence for Scientific Discovery. XII, 311 pages. 2005.

Vol. 3343: C. Freksa, M. Knauff, B. Krieg-Brückner, B. Nebel, T. Barkowsky (Eds.), Spatial Cognition IV. XIII, 519 pages. 2005.

Vol. 3339: G.I. Webb, X. Yu (Eds.), AI 2004: Advances in Artificial Intelligence. XXII, 1272 pages. 2004.

Vol. 3336: D. Karagiannis, U. Reimer (Eds.), Practical Aspects of Knowledge Management. X, 523 pages. 2004.

Vol. 3327: Y. Shi, W. Xu, Z. Chen (Eds.), Data Mining and Knowledge Management. XIII, 263 pages. 2005.

Vol. 3315: C. Lemaître, C.A. Reyes, J.A. González (Eds.), Advances in Artificial Intelligence – IBERAMIA 2004. XX, 987 pages. 2004.